THE DEVIL'S FRUIT

MEDICAL ANTHROPOLOGY: HEALTH, INEQUALITY, AND SOCIAL JUSTICE

Series editor: Lenore Manderson

Books in the Medical Anthropology series are concerned with social patterns of and social responses to ill health, disease, and suffering, and how social exclusion and social justice shape health and healing outcomes. The series is designed to reflect the diversity of contemporary medical anthropological research and writing, and will offer scholars a forum to publish work that showcases the theoretical sophistication, methodological soundness, and ethnographic richness of the field.

Books in the series may include studies on the organization and movement of peoples, technologies, and treatments, how inequalities pattern access to these, and how individuals, communities and states respond to various assaults on wellbeing, including from illness, disaster, and violence.

THE DEVIL'S FRUIT

Farmworkers, Health, and Environmental Justice

DVERA I. SAXTON

RUTGERS UNIVERSITY PRESS

New Brunswick, Camden, and Newark, New Jersey, and London

Library of Congress Cataloging-in-Publication Data

Names: Saxton, Dvera I., author.
Title: The devil's fruit : farmworkers, health and environmental
    justice / Dvera I. Saxton.
Other titles: Medical anthropology (New Brunswick, N.J.)
Description: New Brunswick, NJ : Rutgers University Press, [2021] |
    Series: Medical anthropology | Includes bibliographical references and index.
Identifiers: LCCN 2020020470 | ISBN 9780813598611 (paperback) |
    ISBN 9780813598628 (hardcover) | ISBN 9780813598635 (epub) |
    ISBN 9780813598642 (mobi) | ISBN 9780813598659 (pdf)
Subjects: LCSH: Migrant agricultural laborers—Health and Hygiene—
    California. | Pesticides—Health aspects—California. | Pesticides—
    Environmental aspects—California.
Classification: LCC HD1527.C2 S32 2021 | DDC 363.17/9209794—dc23
LC record available at https://lccn.loc.gov/2020020470

A British Cataloging-in-Publication record for this book is available from the
British Library.

♾ The paper used in this publication meets the requirements of the American
National Standard for Information Sciences—Permanence of Paper for Printed
Library Materials, ANSI Z39.48-1992.

www.rutgersuniversitypress.org

Manufactured in the United States of America

In memory of my dad, Ronald L. Saxton.
And dedicated to the people of the Pájaro and
Salinas Valleys.

CONTENTS

CONTENTS

SERIES FOREWORD

LENORE MANDERSON

Medical Anthropology: Health, Inequality, and Social Justice aims to capture the diversity of contemporary medical anthropological research and writing. The beauty of ethnography is its capacity, through storytelling, to make sense of suffering as a social experience, and to set it in context. Central to our focus in this series, therefore, is the way in which social structures, political and economic systems and ideologies shape the likelihood and impact of infections, injuries, bodily ruptures and disease, chronic conditions and disability, treatment and care, social repair, and death.

Health and illness are social facts; the circumstances of the maintenance and loss of health are always and everywhere shaped by structural, local, and global relations. Social formations and relations, culture, economy, and political organization, as much as ecology, shape the variance of illness, disability, and disadvantage. The authors of the monographs in this series are concerned centrally with health and illness, healing practices, and access to care, but in each case they highlight the importance of such differences in context as expressed and experienced at individual, household, and wider levels: health risks and outcomes of social structure and household economy, health systems factors, and national and global politics and economics all shape people's lives. In their accounts of health, inequality, and social justice, the authors move across social circumstances, health conditions, and geography, and their intersections and interactions, to demonstrate how individuals, communities, and states manage assaults on people's health and well-being.

As medical anthropologists have long illustrated, the relationships of social context and health status are complex. In addressing these questions, the authors in this series showcase the theoretical sophistication, methodological rigor, and empirical richness of the field while expanding a map of illness, social interaction, and institutional life to illustrate the effects of material conditions and social meanings in troubling and surprising ways. The books reflect medical anthropology as a constantly changing field of scholarship, drawing on diverse research in residential and virtual communities, clinics and laboratories, emergency care, and public health settings; with service providers, individual healers, and households; and with social bodies, human bodies, and biologies. While medical anthropology once concentrated on systems of healing, particular diseases, and embodied experiences, today the field has expanded to include environmental disaster, war, science, technology, faith, gender-based violence, and forced migration. Curiosity about the body and its vicissitudes remains a pivot of our work, but our concerns are with the location of bodies in social life, and with how social structures, temporal imperatives, and shifting exigencies shape life courses. This dynamic field reflects an ethics of the discipline to address these pressing issues of our time.

Globalization has contributed to the complexity of influences on health outcomes; it (re)produces social and economic relations that institutionalize poverty, unequal conditions of everyday life and work, and environments in which diseases grow or subside. Globalization patterns the movement and relations of peoples, technologies, knowledge, programs, and treatments; it shapes differences in health experience and outcomes across space; it informs and amplifies inequalities at individual and country levels. Global forces and local inequalities compound and constantly load on individuals to have impact on their physical and mental health and on their households and communities. At the same time, as the subtitle of this series indicates, we are concerned with questions of social exclusion and inclusion, social justice, and social repair—again, both globally and in local settings. The books will challenge readers not only to reflect on sickness and suffering, deficit, and despair but also on resistance and restitution—on how people respond to injustices and evade the fault lines that might seem to predetermine life outcomes. The aim is to widen the frame within which we conceptualize embodiment and suffering.

Dvera I. Saxton's *The Devil's Fruit* captures this focus on the social, economic, and political production of inequality and injustice in the United States. The title (*La fruta del diablo* in Spanish) refers to the strawberries that break the backs and spirits of underpaid, unsupported, and unregulated migrant food workers. Hunched over matted rows, these workers plant, weed, and harvest fruit for the growers, grower-shippers, and massive multinational corporations that supply some 30 percent of the world's (super)market. The year-round availability, variety, quality and affordability of these strawberries, and blackberries and raspberries, for consumers in the United States and globally, depends on synthetic fertilizer, toxic pesticides, and poorly paid labor. The result is a production system that is both brutal and potentially lethal.

The Devil's Fruit is unapologetic and inspiring. Drawing on her work in the fertile Pájaro and Salinas Valleys, stretching south from San Jose and along the central coast of California, Saxton describes the horrendous conditions in which people live and work. But in this devilishness there is also positive action. In this book, as in her life's work, Saxton troubles, riles, and chafes at the inequalities in this market system and its indifference toward those who work in it. She writes of her outrage with the system and pays close attention to the working conditions and the cost to people's health, well-being, and survival: cardiometabolic disease, depression and anxiety, long-term degenerative conditions, and stillbirths. She shows also how engaged and activist ethnographic ethics, methods, and approaches can be used to support activist laborers. Saxton argues that her engagement as an activist anthropologist comes from a place of care. Her anthropological practice is through caring labor, sometimes the most prosaic, in people's everyday lives and their work in social movements for environmental health and justice. This is not dramatic, flamboyant or game-changing care work, but sustained and deeply committed support to others in their own fights for justice and in their everyday life.

The Devil's Fruit provides a model for a new anthropology committed to questioning and disrupting the ideologies, policies, systems, and structures that perpetuate inequalities and responding and reacting to the problems with which anthropologists are confronted in the field. Saxton asks, "What forms can activist research in medical anthropology take?" She unfolds the ways in which anthropological approaches, methods, and relationships work to reveal injustices. But she also provides us with a way to meld activism into our work as anthropologists and our everyday lives, and thus captures an ethics of practice in engaging with our research participants, and collaborators. In *The Devil's Fruit*, Saxton offers us sparkling prose, powerful ethnography, and passionate analysis, but also—and most important—provides us with a powerful model for how we might work at a time of profound inequality and upheaval.

ABBREVIATIONS

1-3-D:	1-3-dichloropropene
1,2,3-TCP:	1,2,3-trichloropropane
AFSC:	American Friends Service Committee
CA DPR:	California Department of Pesticide Regulation
CA EPA:	California Environmental Protection Agency
CalPIQ:	California Pesticide Illness Query
CFT:	California Federation of Teachers
CMA:	critical medical anthropology
DACA:	Deferred Action for Childhood Arrivals
DDT:	dichlorodiphenyltrichloroethane
ESL:	English as a second language
EWG:	Environmental Working Group
ICE:	Immigration and Customs Enforcement
IPM:	integrated pest management
MBAO:	Methyl Bromide Alternatives Outreach
NAFTA:	North American Free Trade Agreement
PVFT:	Pájaro Valley Federation of Teachers
SINJA:	Sindicato Independiente de Jornaleros Agrícolas
SSDI:	Social Security Disability Insurance
UFW:	United Farm Workers
U.S. EPA:	U.S. Environmental Protection Agency
USDA:	U.S. Department of Agriculture
WPS:	Agricultural Worker Protection Standard

THE DEVIL'S FRUIT

INTRODUCTION
Becoming an Engaged Activist Ethnographer

> We must act in solidarity with those rendered vulnerable by a pathogenic
> system and strive against the oppressions.
> —Elizabeth Cartwright and Lenore Manderson,
> "Diagnosing the Structure"

> We want our stories to be pedagogical, to teach a better version of what is
> common sense so that a better version of society can be demanded
> —Amrah Salomón J., "Telling to Reclaim, Not to Sell"

I entered a small industrial warehouse space in downtown Watsonville, California. The smells of smoldering sage and copal—an aromatic tree
sap—hung in the air. Murals of figures with brown fists held high and Indigenous and Mexican American artwork decked the walls inside and out. Bicycle
wheel rims and inner tubes hung from the ceiling, and an assortment of tools
spilled out of boxes and repurposed filing cabinets in a semiorganized chaos.

I arrived at the Pájaro and Salinas Valleys on California's Central Coast, about
a hundred miles south of San Francisco, in the summer of 2010. Finding "hidden
farmworkers" (Holmes 2013; Mitchell 1996) required engaging with a number of
different communities. One of these was the Watsonville Brown Berets, whose
meeting place is the warehouse described above. They are a group of young
Chicano/a and Mexican activists, many of whom are also the children and grandchildren of farmworkers. Some do farmwork themselves to make ends meet, to
support sick and elderly relatives, or to fund their college educations. In addition
to seeking participants for my research, I also sought a community to foster my
own health and well-being as I adapted to a new place.

Why would the Watsonville Brown Berets, let alone farmworkers, want to
work with me, a white woman from Pennsylvania on her path to a doctoral degree
in anthropology? As Melissa Checker observed during her fieldwork with African American environmental justice organizers in Georgia, "people don't always
need an anthropologist," but they often do need other things: tutors, mentors,

FIGURE 1. The Watsonville Brown Berets protesting the use of methyl iodide, 2011. (Photo by the author.)

rides, graphic and web designers, list serve managers, grant and press writers, and volunteers to set up tables, collect signatures, or clean up after events (2005, 193). In other instances, folks want a new friend (Powdermaker 1966).

I ended up doing many of the things on this list and more, as the Berets reversed my request for help and connections. A few weeks after my initial visits with them, a young member, Joaquin, asked if I could help with the group's efforts in the campaign against methyl iodide, a highly toxic fumigant that kills pests, from rodents to nematodes, prior to transplanting crops. Fumigants are also employed to kill pests on harvested products before they are imported or exported.

Methyl iodide is carcinogenic and toxic to the respiratory, neurological, endocrine, and reproductive systems, and it is also attributed to miscarriages, stillbirths, and developmental problems (Froines et al. 2013). Growers of strawberries, raspberries, blackberries, lettuce, and cut flowers—products that at the time of my research dominated the landscapes of the Pájaro and Salinas Valleys—were keenly interested in methyl iodide. Their previous fumigant of choice, methyl bromide, was in the process of being phased out globally via the United Nations Montreal Protocol due to the chemical's role in depleting the ozone layer (Gareau 2013; Guthman 2019).

Science and public health data, as well as observations in farmworker communities, demonstrate a number of chronic, detrimental, and potentially deadly health effects linked to soil fumigant and other pesticide exposure (ATSDR 2020; Gemmill et al. 2013). The major ecological transformations induced by the growth and expansion of California's industrial agriculture have had and continue to have profound, irreversible, harmful, and deadly consequences, especially for the state's rural residents, including many im/migrant farmworkers and their families. A striking example of the harm comes in the form of preterm birth, as well as birth defects, stillbirths, and miscarriages; these affect communities with higher rates of exposure to toxic substances through the air they breathe, the water they drink, the dust that accumulates on their bedroom floors, and their proximity to hazardous waste and highways, among other causes (Padula et al. 2018). Many policy makers with firm pro-life stances routinely overlook these toxic assaults on child and maternal health in rural communities buttressed by industrial agriculture. These environmental conditions and outcomes are not natural phenomena, and they are not God's will; nor are they acceptable costs of doing business. They are rooted in toxic policies and unjust social and economic political hierarchies. As geographer James Tyner states, "One's exposure to death is more and more conditioned by one's position in capitalism" (2019, x).

I had been "hailed" (Gilmore 2008) by the Berets and their farmworker kin and neighbors to direct my anthropological attention to these and other human and environmental harms. My decision to get involved came in part from a feeling that our work as anthropologists is more meaningful when we use our "knowledge and skill . . . to benefit those we study who want our assistance"

(Doughty 2012, 78). In *The Devil's Fruit: Farmworkers, Health, and Environmental Justice,* I describe the inhumane and toxic conditions and treatment endured by im/migrant farmworkers. These are literally built into the design of California industrial agribusiness, which I argue fits the definition of what Peter Benson and Stuart Kirsch call a "harm industry"—that is, "capitalist enterprises that are predicated on practices that are destructive or harmful to people and the environment: harm is part and parcel of their normal functioning" (2010, 461).

The Devils' Fruit not only describes the features and facets of this harm industry but also explores activist ethnographic work with farmworkers in response to health and environmental injustices. I argue and demonstrate how and why dealing with devilish—as in deadly, depressing, disabling, and toxic—problems requires intersecting ecosocial, emotional, ethnographic, and activist labors. Through my own work as an activist medical anthropologist, I have found that these caring labors of engaged ethnography take on many forms and go in many different directions. I critically and reflexively describe and analyze the ways that engaged and activist ethnographic methods, frameworks, and ethics align and conflict, yet in various ways help support still ongoing struggles for farmworker health and environmental justice in California. These are issues shared by other agricultural communities in the United States and throughout the world.

Some of the questions I explore and seek to address include the following: What are the different forms that engaged and activist anthropology can take when fundamental things like community and worker health and well-being, exposure to toxic substances, and other environmental injustices are at stake? What methodologies in our ethnographic tool kits have activist potential? What does it feel like to do activist anthropology? In the words of medical anthropologists Hansjörg Dilger, Susann Huschke, and Dominik Mattes, how do our "encounters through research with human well-being and suffering challenge anthropologists and their interlocutors to reflect on their own ethical and moral [and activist] values and position[s]?" (2015, 2). What are the risks and limits of these different approaches and positions? Above all, how and why do anthropologists "care?" And is caring—as a feeling, as social and emotional work, as a political commitment, a research methodology, a moral stance, and an action—enough (Durham 2016; Heidbrink 2018; Scheper-Hughes 1992)?

I hope that readers can relate to the stories and approaches described throughout *The Devil's Fruit* through their own work. I want us to think about how these ongoing and evolving questions shape all sorts of engaged anthropology and other work: How can anthropologists respond and react to problems that are traditionally only documented and analyzed? How can anthropological methods such as following the object or participant observation be reconfigured and mobilized to support social, health, and environmental justice? What kinds of pragmatic (and not so pragmatic) solidarity can anthropologists engage in? What forms can activist research in medical anthropology take?

And how long does it take? Notably, my involvement with and concern about farmworker and environmental health did not cease with the completion of my research project. Upon my arrival to the Pájaro and Salinas Valleys in the summer of 2010, community members were intensively involved in a campaign against a class of pesticides that is of grave concern for farmworker health. Soil fumigants, including the hotly contested methyl iodide, are among the most dangerous of the many toxic pesticides used in California and global agriculture. While farmworker, advocate, and environmental justice groups successfully contributed to the voluntary retraction of methyl iodide from the U.S. market by the parent company Arysta LifeScience in 2012, thousands of other agricultural chemicals with concerning health and ecological effects remain in use on farms all over the world. Routine exposure to toxic pesticides is one among many of the multilayered health hazards and social and environmental injustices that farmworkers endure, from the lingering threat of detention and deportation experienced by undocumented im/migrants and their families to the risks of sustaining permanent and life altering disabilities from workplace injuries. Contemplating life in a "permanently polluted world" (Liboiron, Tironi, and Calvillo 2018) can make this work challenging, but this means our responses as engaged scholars and citizens must be multilayered, dynamic, and creative. I mean for this book to serve as a testament, one of many more that have preceded and those that continue to emerge, as to why harm industries must be critiqued and dismantled so that new arrangements and relationships that are healing and healthful for people and the planet can exist. Alone, *The Devil's Fruit* is but one *granito de arena* (grain of sand) as the Spanish-language metaphor goes, in ongoing conversations about research as activism and activism as research. It builds on over a century of activist anthropology for health and environmental justice, a history that is not shared often enough.

MEDICAL ANTHROPOLOGY AND ACTIVISM: EQUIPMENT FOR LIVING

Activism is perpetually misaligned and misunderstood. When you see or hear the words *activism* or *activist*, what comes to mind? Do you think of a passionate crowd of protesters, armed with signs and banners, yelling chants and marching boldly along city streets? Do you see people chained to buildings or performing other acts of civil disobedience? Do you think of the folks who made them breakfast, or those who managed the logistics of such a massive public demonstration? Do you think of the people who stayed behind afterwards to clean up the detritus? What about the reporters who document and write about these stories?

Do you think of famous historical figures, like César Chávez or Dolores Huerta, who played significant roles in establishing the United Farm Workers (UFW) union and fought for farmworkers' rights in the 1960s and 1970s? Do you think of the thousands of mostly nameless striking workers who joined them and whose lives were at stake (Bardacke 2011), or those who marched with their children? Do you think of their wives and mothers, who made them tacos, tamales, and burritos for lunch in addition to joining them on the picket lines and along march routes hundreds of miles long, like the one in 1966 organized by the UFW that went almost three hundred miles from the grape town of Delano in the Central Valley to the California state capitol in Sacramento?

Do you think only of elected officials who campaign? Do you think of the people who help get them elected, who fetch their coffee, make runs to the print shop, and otherwise take care of them? Do you think of the people who run but lose their races to more dominant opponents?

Do you think of anthropologists? Do you think of yourself, going to school, learning histories, skills, and methods that may enable you to solve problems or help people? Do you think of other people—teachers, mentors, parents and grandparents, aunts and uncles—who have taught you things, informed your values and your sense of moral and social responsibility?

There are so many different kinds of labor, emotion, and motivation that are involved in activism. Not all of them are immediately visible or acknowledged, but they are all vitally important.

I made a conscious decision to get involved in ongoing activist efforts in the Pájaro and Salinas Valleys, and especially those focused on farmworker health and environmental justice. It became difficult for me to isolate the real-time concerns of farmworkers—pesticide exposure, housing insecurity, occupational injuries, workplace mistreatment, im/migration, and economic and social inequalities—from efforts to advocate for safer and more ecological farming practices, affordable housing, inclusive and quality health care, educational equity for youth, and im/migrants' rights. These issues are intersectional and multilayered.

Empirical and methodological frameworks that isolated variables from one another literally made no sense to me. They certainly did not capture the nuances or emotional weight of farmworkers' stories and lived experiences or what I was observing and feeling myself.

Activist redirections in my work were inspired by *compañerísmo* (Pine 2013; Scheper-Hughes 1995)—that is, friendships and working partnerships with farmworkers, their families, and other community members, such as students and teachers, who were related to farmworkers or otherwise involved in their lives. What Andrew Jolivette (2015) and others refer to as "research justice" often starts through these kinds of friendships "between people who know and care about one another" (L. Martin 2015, 40) and who are willing to help one another throughout life. I asked farmworkers questions, and the more time we spent together, they and their children started asking me questions: What may have caused this rash on my skin? How do I get into and do well in college? Where can I find food for my family while I'm out of work over the winter? I did my best to address these questions, but some of them I could not answer.

We exchanged stories, food, and advice. We continue even now to check in with one another through phone calls, text messenger chats, and occasional visits. My *compañeros/as* worry about me as a woman living and working on her own without her parents or family nearby and no spouse or children to keep me company. I worry about my *compañeros/as* when they get sick or injured, encounter legal troubles, struggle in school, endure unspeakable traumas, and face the consequences of new harmful policies including increased policing and immigration raids in their communities.

These commitments and bonds made me more attentive to the multiple power dynamics and multilayered inequalities and disparities affecting the everyday lives of farmworkers and their families (Saxton 2013). Direct involvement, engagement, and activism are not just a question of what kind of anthropology I want to do or what theory I will apply. I cannot separate my ethnographic approach from my own personal and professional ethics, values, and relationships (Hale 2006). After all, we are not just anthropologists; we are humans too. Activist research became a way for me to express and enact care and concern for the people and communities I work with—now including my students at Fresno State University.

As students, what we learn in the classroom sometimes sends confusing messages about what anthropology is, can be, or should be (F. V. Harrison 2010, 6–7; Kehoe and Doughty 2012). It took me a while to see caring as an asset to research. Care is often defined in positive ways when we express it through our interpersonal relationships. However, as social scientists, we may be encouraged to segregate our personal and professional lives, our caring sides from our empirical inclinations as scholars and researchers. Roy D'Andrade (1995) has argued that engagement in anthropology muddies the validity and objectivity of ethnographic research. Neither anthropology, nor any other social, life, or physical science, can be truly objective. Power and inequality shape the lives of the communities we work with, and they shape our lives and work, as anthropologists and humans, activists and otherwise (González and Stryker 2014; Hale 2008; Nader 1972; Pels 2014).

Our anthropological ancestors did empirical social science work, developed deep and empathetic relationships with research participants, and sometimes participated in activism and advocacy efforts (Lamphere 2018; Low and Merry 2010; Nader 2017; Price 2004; Silverman 2007). Many histories of activist anthropologists are partially or totally obscured, devalued, or selectively forgotten (Doughty 2012; Price 2004). They're akin to the "black sheep" in our families that no one talks about. Anthropology's "bad ancestors," to borrow from Indigenous ethnic studies scholar-activist Amrah Salomón J. (2015, 186), did activism that was not limited to the extraordinary, hypervisible, loud, and dramatic.

They are our intellectual kin who not only documented oppressions and inequalities but actively challenged them through their work.[1]

The truth is, activism is a significant part of the history of anthropology and, more specifically, the subdiscipline of medical anthropology, which is concerned with issues of health, illness, and healing cross-culturally. Generations of anthropologists have used their knowledge, tools, perspectives, positions, and privileges to "[engage, document, promote and support] cultural diversity, social justice and environmental sustainability," even against formidable odds (Veteto and Lockyear 2015, 359; see also Kehoe and Doughty 2012). Critical medical anthropology (CMA) approaches to research formation, methods, data collection, and analysis in particular have significant activist potential. Questions posed by CMA-engaged scholars have included: How do we diagnose social problems and structures (Cartwright and Manderson 2011; Foley 1999; Kleinman and Benson 2006; Quesada, Hart, and Bourgois 2011)? Should we wait for history to be made, or take affirmative stances on issues that we know jeopardize human rights and welfare (Sanford 2006; Scheper-Hughes 1990)? How can we intervene more directly as mediators or brokers, or more affirmatively as activists, in clinical and social work spaces (Kline 2010; Scheper-Hughes 1990)? How can we make reflexivity and engagement less self-centered and more politically productive (Boyer 2015) for health and environmental justice efforts? How else can we use ethnographic methods and relationships to imagine and create alternative socially and ecologically just futures (Burke and Shear 2014), and an anthropology that is "politically committed and morally engaged" (Scheper-Hughes 1995, 410)?

What would happen if we do nothing (Unterberger 2009, 11) in situations of injustice and harm, resigning ourselves to be voyeurs of oppression, violence, human suffering, and ecological destruction? The nineteenth-century physician-anthropologist Rudolf Virchow could not sit by and do nothing. He documented how food shortages exacerbated a rapidly rising death toll amid a typhoid epidemic afflicting impoverished peasants in Upper Silesia, then part of the Prussian Empire (Taylor and Rieger 1985). His positioning as a medical doctor, anthropologist, and outspoken political activist helped Virchow understand that disease and illness could not be assessed or addressed in isolation from the broader contexts and forces shaping everyday lives. Context for Virchow included things like the environment, food, culture, kinship, work and home life (Ackerknecht 1953)—variables that have continued significance in the lives and well-being of twenty-first-century farmworkers.

Virchow did not stop at writing down what he saw. He also publicly challenged German state and church leaders to do something about the grave social inequalities he observed. Virchow knew that reducing poverty and food insecurity were the only viable means of "curing" typhoid. This proposal stood in stark contrast to the endemic scapegoating of the poor for their own sickness and suffering and political interventions that did nothing more than haphazardly distribute medicines. Hunger and poverty made people more structurally vulnerable (Quesada, Hart, and Bourgois 2011) to succumbing to diseases like typhoid.

This radical and still very relevant proposal for *social medicine* explicitly seeks to address health inequalities and disparities through systemic and political changes. Social medicine resonates strongly with social epidemiologist Nancy Krieger's ecosocial approach, which challenges more linear and stagnant models of cause and effect in public health. Krieger (1994, 2001) puts forth a fractal metaphor to help us visualize and respond to the inherent interconnectedness of different life forms and ways of life, including the roles of power and inequality. Such holism is also a central concern of contemporary CMA, which aims "not simply to understand but to change culturally inappropriate, oppressive, and exploitative patterns in the health arena and beyond [and to see] commitment to changes as fundamental to [anthropology]" (Singer 1995, 81; see also Singer 1986; and Singer and Castro 2004). CMA sees the relationships between ill health and global inequalities, based in our exploitative and extractive capitalist system, as inextricable from one another (Baer, Singer, and Susser 2003; Singer 1986, 128). In *The Devil's Fruit*, this includes efforts to address the very individuals, assumptions, and systems that create and perpetuate harm, including California industrial agriculture.

Thus, engaged and activist anthropology is not one thing but many ways of doing ethnography *and* being human. There is no one-size-fits-all approach, as "specific meanings are shaped by the contexts from which . . . dilemmas emerge" (Clarke 2010, S311). It takes place within and beyond universities and other institutions, and across all four of anthropology's subfields (cultural, archeological, biophysical, and linguistic; Low and Merry 2010, S204). Some have come to appreciate its empirical value in producing rich data that are inaccessible through other more conventional means. As Charles Hale notes, "to align oneself with a political struggle while carrying out research on issues related to that struggle is to occupy a space of profoundly generative scholarly understanding" (2006, 98). Anthropologists become activists on account of their personal beliefs and their emotional reactions to fieldwork, professional training, job conditions, ethical concerns, community requests, and/or social scientific curiosity (Bourgois 2010; Dilger, Huschke, and Mattes 2015; Hale 2006; Harrison 2010; Johnston 2001a, 2010; Nader 1972; Scheper-Hughes 1990, 1995; Veteto and Lockyear 2015). Conversely, some come into anthropology through previous activist work (Powdermaker 1966; Nader 1972; Reese 2019; Schuller 2010). In my own experience, the methods of anthropology and activism proved to be cross-fertile grounds, as research, community service, caring and emotional labors, and organizing work reinforced one another. These intersections surfaced, both intentionally and in unexpected ways, through my relationships with farmworkers and my involvement in issues affecting their everyday lives. This includes still ongoing efforts to achieve health care for all and long-standing environmental justice movements against toxic pesticides.

Some anthropologists warn of the challenges that activist research entails, such as the emotional tolls of being involved or the barriers to accessing participants in positions of power. These are very real, but not significantly different from the challenges encountered in more traditional kinds of ethnographic fieldwork, wherein anthropologists struggled to get used to life far away from home and to develop trusting relationships with apprehensive strangers (Nader 1972; Powdermaker 1966). I found that my linked roles as a researcher-activist created unique opportunities for engagement and analysis up and down the agricultural hierarchy, with farmworkers, young people and students, teachers, health care providers, and rural residents, as well as agricultural extension agents, pesticide lobbyists, advocates, and growers. I also found that activism was a way for me to channel the challenging emotions I experienced during fieldwork—namely my "indignity" (Nader 1972), frustration, and feelings of helplessness and hopelessness (Duncan 2018) at witnessing the ecologically and socially toxic suffering and injustices experienced by farmworker *compañeros/as*.

In these many ways, the boundaries between our work and personal lives frequently blur in engagement and activism (Unterberger 2009). Engaged and activist anthropologists often shift gears to adapt to varying conditions in the field and to respond to concerns that come up suddenly or unexpectedly. Being flexible aligns with arguments for mixing methods in ethnographic research to improve data quality and diversity. Multiple and intertwined forms of engagement in research can also strengthen activist efforts (Hale 2006; Low and Merry 2010, S207). What I've elsewhere called "ethnographic movement methods" (Saxton 2015b) encompasses this ever-fluid yet intentional melding of ethnographic and activist methods through a perpetual shifting between roles and positions and the commingling or blurring (Unterberger 2009) of different genres of research, activism, and practice (Heyman 2011; Lamphere 2018; Low and Merry 2010). Ethnographic movement methods also reassert the urgent need and activist call to *move*: to not only observe and document, but to react and respond to communities' social, health, and ecological concerns in more mutual, cooperative, and intersectional ways (Besteman 2015, 260; Saxton 2015b).

Building on the metaphors of motion, emotions, and social movement, Salomón J. uses the clever term *piegogia*—merging *pie* (foot) with *gogia* (study or practice) to describe "that form of learning, resulting from action and experimentation, that gets us to move our feet first and theorize about it later" with every step (*en pie*) we take (2015, 188). The Zapatistas, a pan-Indigenous political revolutionary activist group based in Chiapas, México, invoked *piegogia* to describe the learning by doing of their decades-long organizing efforts for human rights, dignity, and environmental justice. I offer a similar perspective, that engaged ethnography necessitates "participatory (re)action" (Maskens and

Blanes 2013), or conscious and consistent response and movement, rather than a more passive and reserved participatory observation purely for the sake of research and data collection and analysis. Recent ethnographic works demonstrate that activism is ever more urgent as we face ongoing, evolving, and emergent social and planetary challenges (Lamphere 2018, 65; Low and Merry 2010, S204–S205).

ANTHROPOLOGY, FARMWORKERS, AND ENVIRONMENTAL HEALTH AND JUSTICE

There are many similarities between activism and ethnographic methods, as "our core methodologies most resemble that of grassroots activism: participation, holistic listening, and a humanistic approach to caring, understanding, and working with real people" (Schuller 2010, 43). Within the subdiscipline of medical anthropology, concern about the health and ecological consequences of our pathogenic global food system has evoked diverse forms of ethnographic engagement and solidarity for farmworker and environmental health and justice. Throughout *The Devil's Fruit*, I aim to demonstrate how CMA and ecosocial approaches align with or become forms of activism in and of themselves, including environmental justice organizing.

Environmental justice movements and activism seek to end (1) disproportionate exposure to toxic substances; (2) unequal protection under laws and policies, and affected communities' ability to participate in decision-making processes; and (3) inequitable access to resources like clean water and air, land, food, safe housing, and recreational space (as well as other things that keep us happy and healthy) for people of color in the United States and globally (Berkey 2017, citing Bryant 1995, 5; Brulle and Pellow 2006). Environmental justice efforts are not limited to the work of addressing issues in the material world; they are also imaginative, entailing efforts to create, build, and sustain communities in which all people have inalienable rights to health, safety, dignity, and a future. Environmental health and justice issues are of special concern to workers in the global food system, which features long-entrenched racial and ethnic (Cartwright 2011; Holmes 2007, 2013) and gendered (Barndt 2008; Castañeda and Zavella 2003; Zavella 2011) divisions of labor. Im/migrant farm and food workers are at heightened risk for occupational hazards and exposure to toxic substances, loss of livelihood and cultural identities, intergenerational poverty, injuries and disabilities, chronic stress and disease, involuntary displacement, and death (Holmes 2013; Horton 2016b; A. A. López 2007; Smith-Nonini 2011; Sangaramoorthy 2019; Saxton and Stuesse 2018; Stuesse 2016, 2018; Unterberger 2018).

Many anthropologists have supported or participated directly in environmental health and justice struggles within im/migrant and farmworker communities. Interdisciplinary collaborations among medical anthropologists; farmworkers;

worker, community, and environmental justice groups; and others have built, supported, and sustained all kinds of things. These include the creation of ephemeral materials like comic books, films, and other educational interventions to raise awareness about the risks and hazards of pesticide exposure for farmworkers and their families (Flocks and Monaghan 2003, 7; Quandt et al. 2001, 92–93). Anthropologists have also supported farm labor organizing efforts (Daria 2019; Griffith 2009; Smith-Nonini 2011; Stuesse 2016; Williams 1975) to secure more rights and protections for workers that would improve their overall quality of life. Ethnographic data and anthropological critiques and analyses have been applied to change federal, state, and worksite policies and practices that endanger the lives of farmworkers, such as heat and pesticide exposure, housing conditions, and access to health care (Arcury, Quandt, and Russell 2002 Flocks and Monaghan 2003; Horton 2016b; Lamphere 2018; Monaghan 2011; Rao et al. 2007). Longer-term relationships between communities and anthropologists have allowed the witnessing of the health and social tolls that transitions to pesticide-intensive export-oriented agriculture have for people in the countries where many im/migrant farmworkers hail from, including Guatemala (Dowdall and Klotz 2014; Fischer and Benson 2006), Honduras (Phillips 2010), México (Guillette et al. 1998; A. A. López 2007; Wright 2005), and Nicaragua (Bohme 2015b). Combined, this research demonstrates troubling patterns about environmental and health injustices and the hidden costs of industrial agriculture as a harm industry that are global in scope and severity.

Anthropologists have also long participated in public education endeavors, inside and outside classrooms and universities in support of food and farmworkers and in solidarity with different occupational, health, and environmental justice movements. An example of this is anthropologist Angela Stuesse working with the Mississippi Poultry Workers' Center. There she helped develop a bilingual curriculum to support Black and Latinx workers in learning each other's histories, struggles, and languages, along with information about workplace abuses and workers' rights (2016). Similarly, during Nolan Kline's (2019) research on policing and im/migrant health in Georgia, he acted in an LGBTQIA+ im/migrants' rights group's *teatro popular* productions: Spanish-language, improvisational, community-produced plays aimed at helping im/migrants navigate the many day-to-day challenges they face, such as discrimination at work or what to do during immigration enforcement raids. Such efforts can help assuage racial and ethnic tensions, between different groups of workers, and with anthropologists and other outsiders and may foster more collaboration and leadership among workers from different backgrounds. Other examples of engaged scholarship for health and environmental justice include think pieces, articles, and interviews authored by or featuring anthropologists who share their critical perspectives about contemporary issues affecting im/migrant and/or farmworking communities (Holt-Gimenez 2017b; Saxton 2011, 2012; Stuesse, Coleman, and

Horton 2016). Shaping public attitudes and offering cultural critiques doesn't just happen through collaborations with community groups or professional writing (Martin 2013), and education isn't limited to formal classroom settings. An explicit example of this is environmental and medical anthropologist Brian McKenna's roles as an anthropologist and whistleblower. In the late 1990s McKenna was hired to do a statewide environmental health assessment for the Michigan Department of Health and Human Services' Environmental Health Bureau. While preparing the first report, which described disturbing findings about widespread air and water quality concerns and exposure to toxic substances that were clearly linked to industrial and corporate activity, the state agency blocked McKenna's publication. Still, moved by a sense of moral and civic responsibility, he found other ways to share his findings through collaborations with environmental justice organizations and by writing articles for legal journals, community forums, and websites; press releases; and newspaper columns (McKenna 2010).[2]

Another important part of activist anthropology is recognizing the limits and constraints of our efforts and approaches and not leaning too heavily on things that seem pragmatic, practical, or realistic. Throughout my work for *The Devil's Fruit*, I experienced a still ongoing uneasiness between what physician-anthropologist Paul Farmer (2003, 2004) calls "pragmatic" or practical acts of solidarity and the not-so-practical or convenient things that activism sometimes involves. This includes the gut-wrenching labor of caring and the long-term commitment required to effect and sustain structural changes or at the very least keep pushing. I explore these tensions throughout this book, but they remain unresolved for me as a professional engaged in research and social change work and as a human being caught up in the emotionally fraught work of activism and caring about *compañeros/as*.

It is exciting to see new and emerging work in anthropology taking explicitly activist stances. But before delving into the stuff of *The Devil's Fruit*, I would like to reflect on my activist roots and history, along with those of my father and my graduate mentor. I do this to emphasize the ethical importance of acknowledging how who we are (as people and as researchers) and where we come from (geographically, culturally, and intellectually) affects how and why we engage with people and places in our research (Davis 2006; F. Harrison 2010; Kovach 2009; Warren 2006). I also want readers to think about their own histories, positionalities, privileges, and research and social relationships, in ways that help us find reasons to live, work, and organize together across cultures, geographies, languages, occupations, and communities. It is not always easy or straightforward, especially given the stark inequities and violences experienced across different geographies, ecologies, races and ethnicities, genders and sexualities, classes, and citizenship statuses. I feel strongly that such reflections and groundings are necessary preconditions to engaged and activist work of all kinds, building on calls for anthropologists to be more politically reflexive about their roles and relationships in the field (Boyer 2015; Clifford and Marcus 1986; F. Harrison 2010; Kovach 2009).

ACTIVIST GENEALOGIES

During my fieldwork in the Pájaro and Salinas Valleys I would call my dad, Ronald L. Saxton, back home in Pennsylvania, three thousand miles away. I would also regularly email my adviser, Brett Williams, in Washington, DC, where I attended graduate school at American University. I described the disturbing things I observed, heard, felt, and documented. These included abominable housing conditions for im/migrant farmworkers and California laws that sometimes looked good on paper but did not always have the intended protective affects. The people I met and befriended faced seemingly insurmountable challenges accessing health care and navigating other vital social services. They suffered debilitating illnesses and injuries that also brought on chronic bouts of depression and loss of self-worth. On top of that, they were routinely exposed to some of the most toxic pesticides, both at work in the fields and in their homes along the fence lines.

Those phone calls and visits home were often tearful. They also brought up many memories for my dad and Brett that helped me process and cope with what I was experiencing. Both of them had been involved, in different ways, with advocacy, community service, and organizing with farmworkers in two different parts of the United States in the 1960s and 1970s. The legacies of their work and that of other engaged scholars, community organizers, and everyday people, including farmworkers and their families, shaped and continues to inspire my commitment to activist anthropology. Stitching together our ecosocial memories enabled me to see continuities and changes in the status and social welfare of farmworkers, to contemplate the conscious and unconscious ways that my kin and intellectual roots inform my thinking and doing. I want *The Devil's Fruit* to inspire reflections on how one's varied life experiences can be mobilized, irrespective of one's discipline, profession, or context.

ACTIVISM IN THE LIFE COURSE

My dad once told me a story from his adolescence. In the 1940s he heard a knock at the door of his family's northeastern Pennsylvania farmhouse. There stood a young Black man, asking politely for a bucket of water. My dad, who described his preteen self as "young and naive," obliged, and also offered a few slices of bread "for the baby."

Peering out the front door to the roadside, he saw a pickup truck full of migrant African American farmworkers, ranging in age from infants to the elderly, anxiously awaiting a drink. They likely had driven up from the southern states following the harvests up and down the East Coast, en route to the Finger Lakes of upstate New York to harvest summer peaches and then fall cabbages and apples. The man brought the bucket back to my dad, thanked him, and then drove onward, the fully loaded vehicle rattling up the county road.

It is likely that a combination of naïveté, curiosity, and Christian values of charity guided my dad's actions. It was certainly not an activist response to the systemic racism that enabled his parents, the descendants of seventeenth-century British and Dutch settlers in New England, to eventually own their own home and land for dairy farming, even after enduring hardships during the Great Depression. My paternal grandparents had college degrees. My grandmother worked as a schoolteacher, cushioning her family's modest farm income. Having land meant that my then teenage father could grow a garden to supplement his mother's home cooking, along with plenty of milk.

Meanwhile, migrant African American farmworkers worked low-wage harvesting jobs with their entire families in tow. They often lived in informal camps or decrepit on-farm housing (Bloom 2011), including sheds, shacks, barns, and animal shelters. They rarely owned land. Even with a college education, getting a professional job proved difficult in the Jim Crow South as well as the so-called Progressive North. Systemic racism controlled one's destiny.

In the late 1940s my dad left the farm to study chemical engineering. Eventually he earned a doctorate and landed a job with DuPont (the company that has since had many mergers with larger firms like the Dow Chemical Company) during the plastics boom. He and his first wife and their children (my half siblings) settled in southeastern Pennsylvania, seeking out a rural farmhouse life with good public schools. Their growing family became active in the London Grove Quaker Meeting. The Quakers, or American Friends, are known for their pacifist resistance to their own religious persecution in Europe and to war and violence in general, from the Revolutionary War to the endless wars of the present era. In the 1800s many Quakers participated in the abolitionist movement and hosted stops on the Underground Railroad in their basements, crawl spaces, barns, and churches, the remains of which represent a semihidden historical archeology of activism in the northeastern United States.

Quaker settlers also founded the beginnings of Pennsylvania's mushroom industry, first employing Italian immigrants from the neighboring stone quarries as farmworkers. Eventually some Italians became farmers and landed elites. Quaker and Italian namesakes grace packages of mushrooms even to the present day. Mushroom growers established interlocking compost, growing, and marketing enterprises, the largest of which are now vertically integrated. Chester County fresh, dried, organic, pickled, and canned mushrooms are shipped all over the United States and beyond, making the region the self-proclaimed Mushroom Capitol of the World. The industry has been built upon the labors of successive generations of im/migrant and low-wage workers. Throughout the twentieth and into the twenty-first centuries, African Americans, Italians, Mexicans, and Puerto Ricans, and most recently Central Americans, have harvested, packed, and processed Pennsylvania's mushrooms (V. Garcia 1997, 2005).

Through their involvement in the Quaker community, my dad and his family got involved in some of the civil rights and racial equality projects of the American Friends Service Committee (AFSC) in the 1960s and 1970s. This included the founding of an affordable community day-care center in what had once been a local bar. The former Tic Toc watering hole became the Tick Tock Early Learning Center, and provided preschool and childcare to low-income residents, including Black, Mexican, and Puerto Rican farmworker children. My dad helped fundraise, secure, and remodel the building. Some of his children attended Tick Tock, made friends, and picked up some Spanish, as evidenced by the name of the family barn cat, Mamacita. Tick Tock still stands today, over fifty years later, and continues to serve hundreds of children from farmworker, im/migrant, and low-income families.

My dad was also involved with a local chapter of Self-Help Housing, a nationwide project of the AFSC. In collaboration with Black, Mexican, and Puerto Rican communities and church groups, he had learned construction and woodworking skills. Alongside other peers knowledgeable in plumbing, electrical, construction, and community outreach, they helped built affordable housing for families. These folks included several people who were employed in some way by the mushroom industry. They had previously lived in slum apartments equipped with faulty wiring, decrepit trailers with no septic or sewage hookups, or in the drafty uninsulated barns and shacks of their employers among animals and stacks of hay. Some of these dwellings were owned by the landowners and mushroom farmers who attended the same Quaker meetings as my dad and his family. Suffice it to say, the efforts of Self-Help Housing were not always embraced by all community or meeting members.

In the 1970s my graduate mentor and adviser, Brett Williams, began her own anthropological dissertation research on migrant farmworkers. She followed workers along the "circuit," where Chicanos/as and Mexicans traveled each summer from the Rio Grande valley in Texas to a more northern Midwestern town. These farmworkers harvested fruits and vegetables for a then booming canning industry. Brett lived somewhere outside the town of Prairie Junction (a pseudonym). First and foremost, she made friends and acquaintances with farmworkers and their children, community teachers, lawyers, doctors, clergy, and others. Her entry and legitimacy in Prairie Junction hinged on her taking on more socially accepted roles as teacher and social worker, versus researcher and activist. Otherwise, the guards stationed at the labor camps would not have granted her entry. She taught arts and crafts to farmworker children, English to adults, and provided interpretation for farmworkers at their various appointments.

As she described it, she was always "on call" (Williams 1975,10). I can certainly relate to this in my own work. But the community also asked her to do more: to use her skills, social networks, and capital to fundraise for them, and to help them get out of the fields and into other—better paying and less physically taxing— jobs. Eventually they rallied Brett's assistance to create a halfway house for homeless farmworker families just outside Prairie Junction, and she lived in the house to help run it for a time. She also helped organize a lettuce boycott with migrants back in Texas, her home state, as part of the broader efforts of the UFW. Through an expansive network of volunteers and activists, consumers nationwide were urged not to buy lettuce marketed by companies that had refused to sign union contracts with farmworkers (Neuberger 2013). These coincided with table and wine grape boycotts.

Brett's organizing enhanced her research, and her research and friends from the field also shaped her organizing endeavors. Her work helped deflect racial and gender stereotypes about Mexican im/migrants. For example, she framed the gendered labor of Mexican wives and mothers who prepared tamales for their husbands as a critical form of care work that kept families connected and nour- ished in economically and socially oppressive circumstances (Williams 1975).

Forty, fifty, and sixty years later, driving along the rural roads and town byways during my visits home to Pennsylvania, my dad would point out spots on the green rolling landscape to me. There, decrepit barns, shacks, and trailers that housed farmworkers had once stood. Some are still standing, housing new gen- erations of farmworkers.

Today, im/migrant farmworkers continue to face housing struggles along with intersecting and commingling health, social, and political challenges. From Pennsylvania to California and beyond, farmworkers' low wages are nowhere near enough to cover ever rising housing expenses, as well as the increasingly prohibi- tive costs of food, medicine, clothing, and other basic necessities. People cram into single-family homes, trailers, apartments, and employer-owned labor camps that are often in terrible condition (Benson 2008a; Garcia 2005; Holmes 2013). Reading news of immigration raids in both places, and everywhere in between, wrenches my gut. Im/migrant farmworkers are not alone in these struggles for basic needs and rights (Sangaramoorthy 2019). Too often, instead of seeing this as a commonality or an area of shared concern, im/migrants often become the targets of misplaced blame, intense hostility, and criminalization.

Southeastern Pennsylvania is where I grew up and attended public school alongside the children of farmworkers in the 1980s and 1990s. I remember a brown-skinned girl, a "new kid," sitting alone amid the tiny chairs and desks lined up in my fourth grade classroom. Our teacher explained that Marisol *only* spoke Spanish. Marisol rarely looked up from her desk. Occasionally she would be whisked away by another teacher to an English as a second language (ESL) classroom.

After a few months, I never saw Marisol again. Likely, her family moved on—to another job, to a different school district or state, or back to Guanajuáto, México, where many Mexican im/migrants in southeastern Pennsylvania hailed from at that time (Garcia 1997).

I wanted to learn Spanish to communicate with my classmates. In junior high, I remember trading homework with a new friend, Yesenia. She would correct my Spanish, and I helped her with her English-language assignments. In high school I became a tutor in the ESL classroom to earn community service credits toward my graduation. I got my first taste of pozole made by Gloria, a recently arrived student. Her rich dish featured hominy and pork, pigs' feet and many other assorted pig parts, served with all of the fixings—chopped cabbage, onions, cilantro, chiles, and tortilla chips.

When I started college in the early 2000s, against the backdrop of the 9/11 attacks, a series of endless (and ongoing) wars instigated great shifts in how the United States governed its citizens and organized its institutions, including how it handled im/migration. I thought about where I grew up and how these changes affected my former classmates and their families. In my anthropology, Spanish language, and history classes, I started to learn more about the "Peoples and Cultures of Latin America" (as one of my classes was titled), as well as the causes of contemporary global im/migration. These were largely rooted in U.S. economic policies and political interventions.

Immigration and migration into the United States from México is by no means a new phenomenon. There is ample evidence of trade between Indigenous peoples from both regions prior to the arrival of the Spanish. Colonization and the violent succession of much of México into the United States did not automatically end transborder human movement and relationships, even amid efforts to eliminate (by murder or deportation) and assimilate (by schooling, English-only policies, forced removal, and segregated settlement) Indigenous peoples and Mexicans. But from the 1980s up to the present day, the shocks and devastations of a succession of free trade agreements, including the North American Free Trade Agreement, and the political and economic instability following U.S.-funded wars and violence in Central America and México, has led to more and more individuals, and now entire families, leaving their home communities. Land, as well as health, education, transportation, and communication services, became less and less accessible to communities, as these resources shifted into private and corporate control. Hard-fought environmental, labor, land reform, and economic policies were also being gutted and relaxed to make it easier for foreign companies to do business. Rural economies tanked as the value of people's crops declined with the arrival of cheap processed food and genetically modified corn from the United States (Gálvez 2018; Otero 2018). Export-oriented commercial agriculture, from strawberries and cocktail tomatoes to drug cultivation and sales, along with many extractive mining, logging, dam, and energy projects, are largely owned and operated or at the very least controlled by foreign enterprises. Residents of urban and rural centers in México have benefited little from these transformations, despite promises that economic development would bring jobs and opportunities. The rising costs of living; the stagnant wages in agriculture, service work, and industry; and the plummeting market prices of staple crops like corn and beans made life for my grade school classmates, their parents, and the *compañeros/as* (friends, allies, comrades) who I met later during my research in the 2010s, unlivable.

As many of my *compañeros/as* in the Pájaro and Salinas Valleys, and others I have met and befriended while living and working in California's Central Valley point out, they did not migrate "in search of a better life," as the oft-quoted cliché goes. Instead, they came "to have something to eat." They followed the mushrooms to Pennsylvania, and the strawberries and other crops to California in the hopes of having beans, tortillas, and meat on their own tables. Perhaps they could save a little to send back home to support loved ones or build a house made of concrete blocks. Maybe their children would get a good education in the United States so that they would never have to endure the indignities of working in the fields—nor the violence, trauma, and suffering of extreme poverty, hunger, and displacement—again.

Ultimately, the tin miners June Nash met in Bolivia (1993) and the farmworkers Sarah Horton met in Mendota, California (2016b) all astutely observed that the jobs in the mines and the melon fields are all consuming. The word "mines" in the quote, "we eat the mines and the mines eat us" can be replaced with any number of commodities produced by im/migrant workers: melons, houses, skyscrapers, chickens (Stuesse 2016), crabs (Sangaramoorthy 2019), underwear, microchips, coffee, and tea.

My upbringing, my graduate anthropological training, and the further honing of critical thinking and ethnographic methods, theories, and ethics grounded in race, gender, and social justice made abstract intellectual contemplation of these issues feel less than acceptable, let alone just. In our graduate seminars we read classic ethnographies and theory alongside radical activist scholarship. All the while, we constantly contemplated these questions: How has anthropology been used, past and present? What have been the ethical consequences, intentional and unintentional, of anthropological research? What are the uses of anthropology to those being researched? Although I did not anticipate it, I found that when I arrived to the Pájaro and Salinas Valleys, and later the Central Valley, some of these questions would, in various ways, be turned back on me: What made you interested in farmworkers? Why are you living so far away from your family? Why do you care? How can you stand to carry on with your work when it is so heartbreaking and depressing?

Sharing these memories and stories is my attempt at being transparent about where I come from socially, culturally, politically, and intellectually and what my interests and intentions are in working with people and in places with geographic, cultural, racial and ethnic, and class backgrounds different from my own. I am a third-generation college student and a second-generation PhD. My whiteness, social, class, U.S. citizenship, doctoral education, employment in academia, and mostly able body are significant sources of privilege. While not all readers will possess these specific privileges, what we can all consider is how to mobilize our access to education and other resources, skills, life experiences, and training to work in solidarity with our families, neighbors, and communities to resist harm and foster health and healing. In sharing the history of my regional, kin, intellectual, and activist roots, I want to be accountable, to the extent possible, to these questions that farmworker *compañeros/as*, students, fellow scholars, and others pose to me in our relationships with one another.

As I started writing this book, I was struck by the similarities between these three different generations of observations, and community service and organizing with im/migrant farmworkers: my dad's, Brett Williams's, and my own. I've shared many more bowls of pozole since my first with Gloria in 2000: pork or chicken, green or red. Perhaps the most memorable was consumed with a Mixtec family at an informal farm labor camp. The broth was clear, dotted with hominy and shredded chicken, and topped with a swirl of black *mole oaxaqueño*, a rich sauce of chiles, chocolate, and herbs cooked over a wood fire, served steaming hot in a Styrofoam bowl with a plastic spoon, and consumed under a lean-to. The cold New Year's Day rain pattered rhythmically on the corrugated tin roof.

Through these exchanges and *encuentros*, these acts of coming together, I listened to more memories and stories. Those of the farmworkers I met and befriended tell us something about the temporal and spatial ubiquity and patterning of harms in the food system. They reveal not only the possibility of nurturing more nuanced understandings of communities that anthropological approaches may help to engender but also the possibility of melding activism into our work and everyday lives, even over a bowl (or two) of pozole.

What motivates our solidarities with people who come from different backgrounds, from different ethnicities, language groups, class statuses, and geographic regions? Why do we do this work, even when it's sometimes dangerous, deeply depressing, hostile, or seemingly hopeless?

My dad did not join the AFSC's Committee on Race Relations out of guilt, a sense of moral charity, or to get into heaven. Brett and I didn't become anthropologists just for the sake of collecting data and doctorates or mere intellectual curiosity. Rather, our commitments come from a strong sense that an injustice to one is an injustice to us all. Addressing and remedying some of those injustices involves, at first, building relationships, and later, needed infrastructures (material, social, and political), in the form of day-care centers and halfway houses, peer-to-peer and neighbor-to-neighbor relationships, more equitable policies, and responsive and convivial networks of support, care, and activism. This has often meant relying less on elected officials to serve the interests of the public; too often they break their promises to communities, and in some cases it is clear that they could not care less about the issues at stake. In other instances, it has meant being unpopular or receiving subtle (and not-so-subtle) threats, or cutting ties with people or groups perpetuating inequality through gatekeeping or a refusal to engage. Not everyone at the Quaker meetings in southeastern Pennsylvania supported the work my dad and his family and colleagues engaged in. Not everyone shared a commitment to racial justice. The camp guards in Prairie Junction would not have allowed Brett in had she not couched her work in the seemingly more politically neutral terms of social work, childcare, and education. I have had hostile altercations with politicians and other powerful figures. I don't keep doing what I do for a paycheck, or because it feels good—because it doesn't always.

Our actions and intentions have consequences, some known and some unforeseen. As people committed to studying, writing, speaking, teaching about, and working against injustices, as students and practitioners of anthropology and other people-centered professions and activities, *how we care* and *how we think about* problems and ethnographic participants, their communities, and the outcomes of our work matter a great deal. We can afford to be more thoughtful about it, and I believe that anthropology provides critical frameworks as well a tool kit, or "equipment for living" (Myerhoff 2007, 17–27), on our planet.

THE ORGANIZATION OF THIS BOOK

The Devil's Fruit continues contemplating and echoing the question of "applying anthropology to [or for] what [or whom]?" (Remy 1976). By sharing the trajectories and development of engaged and activist medical anthropology in North America that I have herein, and my own activist adaptations of ethnographic methods and frameworks in the chapters that follow, I aim to further destigmatize and normalize activism as a critical and urgent way of doing anthropological research and practice and, perhaps more important, *being human* alongside research participants and *compañeros/as*.

Historically, many anthropologists have focused on collecting myths, legends, and creation stories of cultures different from their own, but rarely were these methods or frameworks applied to the cultures of the anthropologists themselves. In Chapter 1, "Engaged Anthropology with Farmworkers: Building Rapport, Busting Myths," I present some of the myths and misconceptions about im/migrants, im/migration, and farmwork that came up during the course of my research. I juxtapose these with the lived experiences of farmworkers I met and befriended in the Pájaro and Salinas Valleys, most of whom came from México. It is estimated that over half of the im/migrant farmworkers currently in the United States are undocumented (USDA ERS 2019). I elaborate on how Heide

Castañeda's (2010) concepts of *im/migrant* and *im/migration* highlight the vulnerabilities and everyday precarious circumstances (Sangaramoorthy 2019) endured by California farmworkers. Myth busting as an activist ethnographic labor builds on anthropology's documentary and empathetic listening skill sets. It is also a significant part of broader efforts to rehumanize people who are branded and classified as less than human by private industries and state powers that profit from the systemic and routine exploitation of im/migrant farmworkers. Readers will learn about the many social and ecological disparities in California's industrial agricultural communities that interact synergistically and produce intersecting health problems for farmworkers and their families.

Another strategy we can mobilize to make the uneven power dynamics, toxic relationships, and human suffering characteristic of harm industries more tangible and actionable is to "follow the object." In Chapter 2, "Strawberries: An (Un) natural History," I engage in a critical commodity chain analysis, elucidating the many contradictory ethical, ecological, and cultural values and practices involved in the production of strawberries and other fruits and vegetables in the Pájaro and Salinas Valleys. Industrial agriculture's dependence on toxic pesticides has especially devastating consequences for farmworker health and perhaps the future of agriculture itself. Social scientists have followed foods and other agricultural commodities all over the world, describing the "social lives of things" (Appadurai 1986, 13), dissecting the unique ecological, technological, cultural, and market and labor conditions and pressures specific to different agricultural products. And they use this information to challenge socially unjust and ecologically toxic practices in agriculture and food production (Friedland 1984, 1994; Friedland, Barton, and Thomas 1981; Guthman 2009; Hartwick 1998). Some studies, for example, have followed the material, sociopolitical, and ecological routes; the unequal social relations; and the health and ecological consequences of pesticides in Bhopal, India (Fortun 2001), organic and locally grown produce in California (Guthman 2004) and upstate New York (Gray 2013), broccoli imported to the United States that is grown in Guatemala (Fischer and Benson 2006), tobacco in North Carolina (Benson 2011), tea in Darjeeling, India (Besky 2013; Sen 2017), and factory-made tortillas on both sides of the Mexican-U.S. border (Bank Muñoz 2008).

Strawberries as a starting point connected me to a place, the Pájaro and Salinas Valleys in California; to a community of im/migrant farmworkers and their families; to a crop with a specific history, ecology, and commercial logic of industrial production; and to activists and allies in antipesticide organizing. Understanding the biology, ecology, history, and culture of the community led to deeper understanding of how the strawberry industry itself is structured, governed, and organized and the ecosocial—or fractal and branching—consequences this has for farmworker health. Indeed, the brutal pace and painful bodily contortions demanded of farmworkers are, at least in part, why some farmworkers call strawberries *la fruta del diablo*, the devil's fruit (Schlosser 2003). A number of toxic pesticides and fertilizers, including the especially dangerous soil fumigants, are routinely used in strawberry and other fruit and vegetable operations. In following the strawberries I came to understand how and why consumer desires for freshness, purity, localness, safety, convenience, fairness, ethics, health, wellness, and pleasure are so alienated from farmworkers' lives.[3]

By merging frameworks from critical medical anthropology with the embodied narratives and grounded insights offered by farmworkers, their teenage and adult children, and community members and activists, I aim to destabilize popular and positive images and imaginaries about strawberries and strawberry fields. In Chapter 3, "Pesticides and Farmworker Health: Toxic Layers, Invisible Harm," I present information about pesticides and farmworker health and challenge the distinctions between acute reactions to pesticides and the multilayered chronic harms they engender throughout the life course. Toxic layering (Swartz et al. 2018) entails the pervasive presence of pesticides both on and off the farm in farmworkers' lives, which are exacerbated by toxic policies and gaps that fail to act on both what is known about the harmful effects of pesticides and what is still uncertain or contested. Anthropological frameworks of syndemics and chronicities attend to the multilayered and long-term consequences of exposure to toxic substances in farmworkers' lives and across generations. Many pesticides act epigenetically, altering the life courses of individuals who have yet to be born. Syndemic and long-term approaches are desperately needed in antitoxics research and activism.

In Chapter 4, "Accompanying Farmworkers," I emphasize how the relationships between anthropologists and community members create opportunities to observe and collect data but, just as important, how these friendships are a form of activism (Jackson 2010; Low and Merry 2010, S208; Powdermaker 1966). The activist medical anthropologist Nancy Scheper-Hughes (1995, 410–411) describes how *compañerísmo*, or working friendships based in solidarity, mutual respect, and reciprocity, challenges the objectification of people as mere research subjects, faceless laborers (Benson 2008a), or people without desires or agency (Loza 2016). These relationships, built on trust over time, foster "shared commitments to visions of social justice or social change"; in other instances they involve the sharing of "housing, food, medicine, automobile[s], and other economic, material, and social resources, both at home and in the field" (Low and Merry 2010, S208; see also Ali 2010; and Stuesse 2016).

In these and other ways, I argue that research can be a form of care work, which contrasts sharply with the formal, philanthropic, employer-provided, and other forms of care that farmworkers experience or receive in institutional settings. The word *acompañimiento* (accompaniment) to indicate a form of mutual aid, solidarity, support, and care for one another, irrespective of life circumstances, and shows up in both the activist and anthropological literatures (Duncan 2018;

Farmer 2003, 2011; Indigenous Action Media 2014; Lynd 2013; Scheper-Hughes 1992, 1995). Formal social service institutions do not always provide this level of intimacy, and sometimes care—as a service or even as an emotional and human gesture—is denied on the basis of one's social or immigration status.[4] Through accompaniment, anthropologists can mobilize our skills as interpreters and navigators of systems and our time and energy as concerned citizens. This creates spaces of "emotional care," but, as Whitney Duncan (2018) notes, such care is also limited by the very structural forces that bring such groups of marginalized people into existence in the first place.

Accompaniment is a kind of "pragmatic solidarity" or everyday acts of care and support (Farmer 2003, 2004, 2011; Holmes 2013), but the word *pragmatic* also has limitations. It implies actions deemed realistic, feasible, or doable in the here and now. Pragmatism in research and solidarity efforts can be conditioned by what Laura Nader (1997) calls "controlling processes" or what KumKum Sangari refers to as "the politics of the possible" (1986). Our social, geographical, political, and temporal locations and positionalities also shape our perceptions of what is real and what is accomplishable or actionable. This conflict—what Paul Kivel calls the tension between "social services or social change" (2017, 129–130)—frequently surfaces in activist research. It also manifests through the embodied and emotional feelings of hopelessness experienced by anthropologists who accompany im/migrants, farmworkers, and other marginalized peoples whose circumstances will not change without significant structural upheavals (Duncan 2018; Heidbrink 2018; Stuesse 2016). I want to suggest that in taking on these roles of accompaniment, advocacy, friendship, *compañerismo*, and deeply politicized care, accompaniment can "disrupt *expected* academic roles and statuses" (Scheper-Hughes 1995, 420, emphasis added) and question the logics of care in other contexts that are long overdue for such transformations. I have come to see the ecosocial and emotional labors of accompaniment as a means of modeling research as care through the human relationships, social support, and structural change that they can potentially engender, even in seemingly hopeless situations. Being there, even when we cannot change systems and circumstances, can have profoundly humanizing effects.

Sometimes activist work feels hopeless due to the longevity, intensity, and multilayered diseases, illnesses, social suffering, and environmental violence endured by im/migrant farmworkers and their families (Duncan 2018; Heidbrink 2018). Yet I still see activist potential in merging ecosocial thinking with environmental health and justice activism and research, and I was inspired by unexpected solidarities I observed during my fieldwork. In particular, public school teachers and the student children and grandchildren of farmworkers expressed deep concerns about the effects of pesticides like methyl iodide on community health. They were also the ones who took the lead in regional antipesticide organizing efforts. In Chapter 5, "Ecosocial Solidarities: Teachers, Students, and Farmworker Families,"

I suture together the epidemiologist Nancy Krieger's ecosocial theory and activist medical anthropologist Adrienne Pine's concept of somatic solidarities. Ecosocial frameworks are well suited to documenting and "diagnosing the structures" (Cartwright and Manderson 2011, 451) affecting and afflicting farmworkers' health and well-being. Ecosocial approaches are also organized around activist questions of accountability and justice, or "who and what is responsible for population patterns of health, disease, and well-being, as manifested in present, past and changing social inequalities?" (Krieger 2001, 668). Farmworkers and their teenage children and grandchildren, public school teachers, and environmental and labor union activists posed these questions and challenges in different ways. Teachers, unions, and youth reacted to these intersecting injustices and toxic risks by linking them to their own occupational and community health and safety concerns about pesticide drift from fields bordering schools and homes. In other words, ecosocial solidarities emerged from "shared *embodiment . . .* and *resistance*" (Pine 2013, 144, emphasis added) to pesticides. This also fostered transoccupational and intergenerational solidarities across racial and ethnic, class, and educational differences among farmworking and nonfarmworking community members.

The affective labors of care and solidarity in environmental justice activism regularly clashed with the dismissive tone from agribusinesses, which sought to discount people's concerns about pesticides and farmworker welfare as overly emotional irrational, hysterical, or psychosomatic. The feelings at witnessing or experiencing the direct embodiment of injustices generate the energies necessary for social change. Geographer Julie Guthman has argued that such conflicts between industries and communities are well rehearsed and performative; each side of the debate seems to rehash the same argument for or against toxics over and over again (2017, 5). She has suggested that environmental justice groups make more efforts to compromise with growers and industry and negotiate with them around intersecting vulnerabilities (Guthman 2017, 16; 2019); however, this overlooks long-standing power imbalances between industry and farmworker communities and neglects the glaring absence of explicit and cohesive solidarity with farmworkers from the agricultural industry. Many California farmers and growers voted for Donald Trump in 2016 and will vote for him again in 2020. The Trump administration has terrorized and deported members of the core im/migrant farm workforce. Labor is also often framed in the industry as a problem requiring a technical fix (Guthman 2019)—use more fumigants to increase yields, recruit more workers from the H2-A guest worker program, resist increased labor and environmental regulations, mechanize—even as many of these continue or exacerbate already serious social disparities.

Teachers' and students' participation in anti-pesticide organizing does not stem from absolutist resistance to the agricultural industry or even a complete rejection of pesticides. Instead their activism comes from a deep place of care and concern for their family members, students, and neighbors. Irrespective of the measurable *effectiveness* of their activism on the governance of pesticides, their *affective* expressions and efforts to imagine alternative futures (Burke and Shear 2014) for food, farming, and work deserve validation and support. In these ways ecosocial solidarities among farmworkers, their children and grandchildren, public school teachers, labor leaders, and other community activists develop through both the physical and emotional dimensions of embodiment. As Max Liboiron, Manuel Tironi, and Nerea Calvillo observe, "activism based in ethics rather than achievement" has value (2018, 331).

In the Conclusion, "Activist Anthropology as Triage," I take stock of some of the frustrations and stresses of being an activist researcher in the era of Trump. While im/migrant and farmworking communities in the United States have long endured marginalization and hostilities, irrespective of who has been president, the explicit intent to harm, venemous decrees, policy changes and rollbacks of the Trump administration are sending shockwaves through the communities where I live, work, and do activist research (Saxton and Stuesse 2018). This makes the tensions between dedicating ourselves to social service and social change work even more fraught (Kivel 2017). I conceptualize the shifting of energies and emotions between supporting people directly, doing the work necessary for health and environmental justice, sustaining long-term social change work, and nurturing and caring for ourselves and one another as a form of activist triage.

In hospitals and emergency rooms, triage mandates that care and resources be diverted to those patients deemed most at risk or in need. In reality, triage does not always work this way. To counterbalance the distress, fatigue, and hope-lessness that being in triage mode often evokes, activist anthropology also has a responsibility to support communities in the imagination and creation of alterna-tive futures and food systems and to go beyond pragmatic responses to problems. One gut reaction from readers may be the temptation to stop eating strawberries altogether, or to shift food dollars to more "ethical" products (Saxton 2015a).[5] While organic products may require fewer or less toxic pesticides, and may slightly assuage our consciences, broader cultural and paradigm shifts in food, farming, health care, and im/migration are needed if we are to see significant improve-ments in the lives of farmworkers. The pact (F. Harrison 2010) cannot end at the tip of our forks.

ADELANTE (ONWARD)

There is a great diversity of twenty-first-century activist anthropology for health and environmental justice demonstrating multiple "way[s] of working" (Ingold 2014, 390) and pushing against the limits of traditional disciplinary scholarship and pragmatic forms of research, activism, and solidarity. *The Devil's Fruit* repre-sents a multipronged ecosocial critique that documents and draws awareness to the invisible harm (Goldstein 2017) posed by toxic pesticides and other injustices afflicting farmworkers. I see activist anthropology as a way of synergistically researching, reacting, and responding—in real time and gradually, over time—to the slew of bedeviling and overwhelming problems we and others face.

Very few students of anthropology and other social, health, and environmental studies, sciences, and caring professions will go on to become doctoral-level researchers, and fewer still will become academics themselves. This does not mean that this book is irrelevant to them. On the contrary, I hope that *The Devil's Fruit* will be a useful resource for readers across many career, life, and activist paths. Now, more than ever, we need engaged anthropologists and researchers along with engaged doctors, nurses, lawyers, accountants, social and public health workers, teachers, farmers, chefs and food entrepreneurs, labor and community organizers, activists, advocates, and other everyday, ordinary people. Our work collectively needs to move beyond documenting and diagnosing and toward "disrupting" the assumptions, systems, and structures that enable and perpetuate inequalities and injustices (Davis 2006), ill health, human suffering, and ecological destruction. Given what is at stake in the world, especially for im/migrant farm and food workers (today, as in the past), we need to take community requests more seriously and consider how our methods can become more integral parts of movements and community efforts to address concerns. I hope readers are roused emotionally and viscerally to develop activist research and responses in their own communities.

1 · ENGAGED ANTHROPOLOGY WITH FARMWORKERS
Building Rapport, Busting Myths

I first met Omar, an Indigenous farmworker from Oaxaca and a speaker of the Triqui language, at a community garden, where he and other *paisanos* (compatriots) and their families participated in a very different kind of farmwork on the weekends. They grew *papalo* (a pungent herb served with tacos or hearty and meaty soups and stews), heirloom dent corn, *frijoles de mayo y junio* (beans that flower in May and June), and special varieties of chile peppers for home consumption.

They were not alone in their efforts. In the small urban yards of farmworker households in Salinas and Watsonville, people clearly took a lot of pride and pleasure in growing their own corn, beans, climbing chayote squash, the edible nopales cacti, herbs, and flowers. People enjoyed growing things, even if all they had was an eighteen-inch-wide strip of soil bordering a small dwelling.

Throughout Salinas, Watsonville, and other agricultural cities and towns in California, farmworkers re-create their own versions of life in San Juan Copala, Oaxaca, or Zamora, Michoácan, through gardening and foodways, patron saint and Mexican Independence Day festivals, shared language, traditional dance and music groups, and the perpetual exchange of *chisme* (news and gossip). In these ways they are not unlike previous generations of im/migrants who have left cultural, linguistic, and other marks on communities throughout the United States.

For about a year, I hung out at the garden, weeding and socializing. One of the gardeners' main advocates, Robin, asked if I could helped organize a dinner banquet fundraiser so the group could buy a tractor and celebrate and share their harvest with others. Triqui women prepared a giant pot of goat stew called *masa*, rich with tender cuts of meat and chewy *tripas* (intestines), accompanied by massive piles of crunchy, one-foot-wide, handmade corn *totopos* (large toasted tortillas) and fresh herbs (papalo and cilantro) from the garden. The group reached

its goal and bought a secondhand tractor to make cultivating the land and form-
ing rows easier.

Helping out with this and many other events and efforts was one among myr-
iad ways I built rapport with im/migrant farmworkers: I attended to their welfare not just by listening to them but through direct involvement in issues they
cared about. Some of my efforts were in the short term, like the tractor fundraiser, while others, like my participation with antipesticide activists and organizations like the Center for Farmworker Families, the nonprofit organization
founded by Dr. Ann López, lasted for the duration of my fieldwork, and sometimes longer.

One Sunday, during a break in the community gardening routine, Omar and I
sat under the shade of a eucalyptus tree to talk about his life, work, family, and
im/migration. He described to me the varying circumstances that led to his
movement—back and forth across the Mexican-U.S. border and through several
U.S. states, sometimes with and sometimes without a visa—in search of desperately needed work.

When Omar was seventeen years old, a labor contractor arrived in his community in Oaxaca, seeking young men to come to the United States to work
through the H-2A temporary agricultural worker program.[1] A select number
of H-2A visas are issued annually to agricultural employers who can prove
that their efforts to find workers locally have failed. At present, there is renewed
interest in expanding the H-2A program and the number of grower applications
for H-2A workers is growing significantly (P. Martin 2017). A significant number of H-2A workers are being brought to the Pájaro and Salinas Valleys, as
growers argue that they cannot find enough workers locally, yet the unemployment and underemployment rates for Monterey County and southern Santa
Cruz County remain high (BW Research Partnership 2018). Latinx im/migrants
may have been pulled to California to work in agriculture, but they are also
pushed out of work through deportations, disabilities, and desperation. Agricultural jobs don't pay enough to live on, and im/migrant workers are often
unable to secure positions in higher-paying professions, including the tech and
service-sector jobs that have started coming into the area. Language and educational barriers are further exacerbated by class inequalities that require people to
work just to survive and support their families (Zavella 2016). In addition, immigration raids, the exorbitant and ever rising costs of living on California's Central
Coast, and the aging and ageist displacement of previous generations of longer-
term resident farmworkers (Stoicheff 2018), are also contributing to the farm
labor shortage. So too has the overall decline in undocumented migration from
México since 2012 (Gonzalez-Barrera 2015). Amid all of these changes, efforts
are also underway to reduce the already paltry and patchy protections and rights
that temporary H-2A workers and resident farmworkers have (Costa 2019).

A common plea of many California growers and farmers is that the costs of state-mandated labor and environmental protections for farmworkers are prohibitive to profitability. These are some of the ways labor shortages are the products of social engineering, not merely the result of people moving on, up, or out.

H-2A or temporary agricultural workers who apply for visas and make it through the federal multiagency approval process and screening receive a short-term work permit along with a contract that ties them to a specific grower. The visa is good for the duration of the harvest, after which workers would be sent back home to their communities. H-2A workers are also supposed to receive transportation from México to the United States and back, and to and from worksites, safe and decent housing, and three meals a day for the duration of their job tenure. This is what the contractor promised Omar and others in the late 1990s, along with a salary of twenty-five dollars a day plus a piece rate for each tree planted. Participation in the H-2A program marked the beginning of Omar's still ongoing im/migrant "farmworkers' journey" (A. López 2007).

In 1998, recently married and with a baby on the way, Omar took the labor scout's offer. Twenty-five dollars a day was much more than he could make work-ing in the tomato fields in the northern Mexican states of Baja California Norte or Sinaloa (Daria 2019; M. López 2011; Wright 2005; Zlolniski 2019), and signifi-cantly more than he would earn doing random construction jobs or through the subsistence farming of corn, beans, and squash at home. The costs of housing, food, children's schooling, health care, fuel, and other basic needs keeps rising throughout México despite the country's place in American popular imagination as an affordable destination for vacations or retirement living, bountiful cheap goods, and lower-cost prescriptions and medical and dental procedures.

From Oaxaca, Omar and his *paisanos* traveled north to Nuevo Leon, Monter-rey, México, on the border with Texas. When they arrived at their destination in Arkansas, they stayed in a run-down motel with six to eight people per room. Working no fewer than six days a week, Omar's crew planted thousands of pine trees as part of a reforestation effort. They received few to no breaks. None of the supervisors spoke Spanish, let alone Triqui.

That twenty-five dollars a day that had sounded so tempting to Omar while he was still in Oaxaca quickly dwindled; the contractors deducted rent, food, rides, and money for work clothing and equipment from all of his checks. Work-ers' wage rates also shifted at the whims of their supervisors. Omar and other crew members were sometimes paid by the hour and other times by the piece rate per box of pine trees planted. Supervisors used these inconsistencies when they wanted to push the crew to work harder and faster to get the job done.

Ultimately, despite his hard work, Omar had little to nothing to send home to his wife and newborn daughter. The stress and pressure of being unable to sup-port his family, combined with homesickness and loneliness, the grueling pace and conditions at work, and the exploitation and abuse he suffered from his supervisors made Omar feel sick. He developed stomach pains and diarrhea that he thinks were due in part to the stresses of the job and being so far away from home with no one to turn to for help.

After two months in Arkansas, Omar and some of his coworkers followed a lead from a fellow *paisano* to head to Florida. Rumor had it there was plenty of work in the tomato fields, which promised a higher piece rate, and more of the supervisors spoke Spanish. The risks of breaking their H-2A contract did not outweigh the possibility that they would make more money to take care of their families. They stayed in Florida through the end of the tomato harvest before returning back to their village in Oaxaca for the winter.

Omar continued to migrate back and forth between Oaxaca and the United States using coyotes (guides), or, increasingly, traffickers who transport im/migrants across the Mexican-U.S. border. When I was doing my research, farmworkers stated that coyotes charged a couple thousand dollars per person per trip. Today the rates are much higher, ranging from seven thousand dollars to upwards of fifteen thousand dollars per person per crossing (Rohrlich 2019). The journey has also become more dangerous and difficult with the growing presence of Customs and Border Patrol agents (Cantú 2018, De Leon 2015). As people are more desperate to get across, they've become more vulnerable to criminal traffickers who extort workers and their families for ever increasing sums, often under threats of violence and rape (Greenfield et al. 2019).

The anthropologist Heide Castañeda offers the concept of im/migrants in order to better capture the variabilities and precarities that mark the life courses of people, like Mexican farmworkers, who move, resettle, and sometimes return to their home communities for work. Im/migration is the crossing and recrossing of national, political, urban versus rural, and social boundaries (Castañeda 2010, 7–8). It encompasses not just geographical movement but major life changes and shifts with respect to people's identities and roles: as gendered individuals, as Mexicans, as Indigenous peoples, and as workers. Im/migration also reflects the uneven and unequal experiences im/migrants have when they are received (or not) in the communities where they temporarily, semipermanently, or permanently settle (Stephen 2007).

For farmworkers, in particular, the fluid concept of im/migration better describes how their livelihoods are contingent upon numerous and intersecting uncertainties (Castañeda 2010). For example, the drought in California from 2013 to 2016 left many farmworkers unemployed and in some cases homeless; more generally, many experience routine seasonal unemployment when crops go out of season. Economic and ecological vulnerabilities and social and political hostilities in home and host communities and at the Mexican-U.S. border also affect people's movement. Some farmworkers migrate between crops and seasons to piece together their incomes, although this practice is also changing. Increasingly, resident farmworkers are being bussed in to agricultural fields from long distances. My acquaintances in California's Central Valley often travel hours to reach worksites arranged by farm labor contractors, who sell their crew's labor to growers. It is rare for farmworkers to be hired directly.

Im/migrant farmworkers' lives in the Pájaro and Salinas Valleys, the purported global berry and salad bowls, are riddled with inequalities, as they are elsewhere in the world. When I conducted most of the research for *The Devil's Fruit* (2010–2013), there were an estimated three million farmworkers in the United States (NCFH 2012), many of whom were born in México. Some lived and worked on their own; many others were accompanied by their families. Many initially found their way across the Mexican-U.S. border to host communities in the United States through extensive networks of family, friends, and *paisanos* who had made similar journeys decades earlier (Zavella 2016). Entire villages, including Omar's, have been uprooted, semipermanently, resettling in commercial and industrial farm towns in the United States, as well as those that have sprung up in places like Culiacán, Sinaloa, and San Quintín, Baja California Norte. Some are able to send remittances to their home communities, as is evidenced by the construction of new homes in villages with significant immigrant exoduses. Others do not realize their goals, as is evidenced by the skeletons of housing structures started but left incomplete.

It is uncertain how many im/migrant farmworkers remain in the United States in 2020 given that net migration from México reached zero in 2012 (Passel, Cohn, and Gonzalez-Barrera 2012).[2] Labor economists Philip Martin and J. Edward Taylor estimated there to be about 2.4 million farmworkers in the U.S., most of whom lived and worked in California (2013, 4). There have been reports of immigrants fleeing the United States for fear of being detained and deported. Nonvoluntary deportations at the hands of U.S. Immigration and Customs Enforcement (ICE) in rural farmworking communities in California and elsewhere have risen dramatically under the presidency of Donald Trump (Grabell and Berkes 2017).

Eventually Omar found decent-paying and steady construction work and a strong Triqui community outside of Atlanta. His wife and daughter joined him, and they grew their family by two more: another daughter and son. Omar wanted to learn English, but none of the adult school programs would accept him. As anthropologists Nolan Kline (2019), Angela Stuesse (2016, 2018), and Alayne Unterberger (2018) have described, Florida, Georgia, Mississippi, and other southern states have fiercely anti-immigrant policies that limit many undocumented immigrants' access to resources and support—from education to health care. The Chicana anthropologist Patricia Zavella (2016) documents how workers affected by the closures of Watsonville's canneries and freezer plants in the late 1980s and 1990s participated in language and job training programs that were supposed to insulate them from the shock of their job losses; however, the programs failed many workers who could not afford not to work while trying to

FIGURE 2. Farmworker im/migration: a map of the Mexican states and the Central, Pájaro, and Salinas Valleys of California. (Image created by Caroline Prioleau.)

learn English and new job skills as adults. Laws that explicitly exclude immigrants and even their U.S. citizen family members from social supports are in the process of expanding nationwide under the Trump administration's policies. Racist discourses that scold newer immigrants for not learning English fail to acknowledge these and other barriers, as described in the stories of Omar and other farmworkers in this chapter.

A significant downturn in the construction industry and housing boom of the mid-2000s made life even more challenging for Omar and once again he and his family returned to Oaxaca. From there they moved to the Pájaro Valley, following the lead of extended kin and friends.[3] Around Watsonville he found steady, year-round, but still arduous and stressful employment as a field irrigation technician, while his wife Aurelia worked seasonally picking raspberries under subcontract for a large grower-shipper. They lived in a single-family home on the outskirts of Watsonville, along with other extended family members. Adults and teens, some as young as twelve or thirteen, pooled resources to be able to afford the rent of twelve hundred dollars a month, even when there was no work in the fields from November through April.

In 2016, shortly after the election of President Donald Trump, Omar and his cousin Benito messaged me, despondent. Benito, in particular, sought to preempt a potentially violent and traumatic encounter with ICE, and considered self-deportation with his entire family, including his younger U.S. citizen children. Ultimately he was undecided: "No se que hacer" (I don't know what to do).

BUSTING MYTHS: IM/MIGRATION AND IM/MIGRANTS

Omar attempted to pursue work in the United States legally through the H-2A temporary worker program. The dire and hostile conditions he faced made it difficult for him to feel human, let alone support his family—the main reason he migrated in the first place. Perhaps he could have reported the contractor for the violations and abuse, but who could he have trusted? When im/migrant workers file a claim or a complaint, they are met with silence or explicit requests that they not return to their jobs: "If you don't like it, then leave!" Where would Omar have gone to make a report? If he knew where to go, how would he get there without his own transportation? How would he communicate if the person on the phone or on the other side of the desk didn't speak Spanish? When he attempted to learn English, a demand often launched at im/migrants to prove their deservingness to exist in the United States, he could not enroll in classes due to laws that explicitly excluded undocumented people from accessing adult and higher education. Ultimately, Omar and his family may never have come to the United States if their lives and livelihoods in Oaxaca had been more livable. He did not come to steal anyone's job or to access free health care or other benefits (things that do not exist in the United States). Instead, contemporary labor im/migration is shaped by inequitable and unjust forces that characterize our global food system.

Assertions that people who want to live and work in the United States should follow legal, social, cultural, and other rules, and peruse legal pathways to residency and citizenship, ignore how im/migration policies have been designed to selectively facilitate the arrival and acceptance of some groups and hinder or prevent them for others. They overlook how different im/migrant groups have been unevenly welcomed in our society based on the xenophobic and racist sentiments and actions of those who hold power and influence. Current U.S. immigration laws and popular beliefs also fail to acknowledge long-standing ties between communities along the Mexican-U.S. border and over a century of dependence on im/migrant workers in many U.S. industries (McWilliams 1966).

This chapter takes on these issues, as well as myths about im/migration and im/migrant farmworkers that came up frequently during the course of my fieldwork. I counter the pervasive myths with my own observations and farmworkers' testimonials about their im/migration stories, work experiences, and daily realities. Our collective consciences and political climate are riddled with deceptive ideas about im/migration and im/migrants that merit deeper ethnographic scrutiny. One of anthropology's traditional research methods, participant observation, involves spending time getting to know communities through casual conversations and everyday interactions and observations. For example, classically trained North American anthropologists of the twentieth century spent years of their lives in communities to collect the myths and legends of different ethnic and cultural groups. This helped them understand how a community thinks, what its members believe, and how these thoughts and beliefs shape behaviors, including rituals and the rhythms and activities of everyday life. Such approaches were rarely applied to question the beliefs and assumptions that guided the academic cultures and Euro-descendent communities from whence anthropologists hailed.

Narratives that demean and dehumanize im/migrants are infused with racism that have been used to justify fewer rights, normalize occupational hazards, cultivate acceptance of lower wages, and excuse the lack of quality care and support, and basic resources like food and housing, afforded to generation after generation of im/migrants. Alternatively, there are myths about Mexican im/migrant farmworkers, some embraced within the community, that deify and romanticize them as superhumans who are naturally suited to farmwork and other physically demanding tasks (Holmes 2007; Horton 2016b). These positive associations are also dangerous to worker health, especially when they naturalize problematic race and ethnic, gender, and citizenship divisions of labor that resign certain groups to the most hazardous jobs (Benson 2008a, 2008b, 2011; Cartwright 2011; Holmes 2007; Horton 2016b; Smith-Nonini 2011) and the most marginal spaces in our society.

Instead of solely focusing on the beliefs and behaviors of farmworkers as drivers for health outcomes, we need to recognize our systems of im/migration and industrial agriculture as interlocking *institutional cultures*, "harm industries" (Benson and Kirsch 2010) that present significant consequences for farmworker health. This allows us to analyze and challenge the social assumptions and logics shaping otherwise taken-for-granted labor, im/migration, and environmental policies and practices that become embodied by farmworkers in the forms of ill health, disease, death, and diminished life opportunities.

Before engagement and activism can happen, we must first recognize im/migrant farmworkers as fellow humans. I came to know Omar and many others through *engaged ethnographic research*: ways of doing research and thinking anthropologically grounded in a deep commitment to the communities where we live and study. Interspersed within my efforts to bust myths are descriptions of farmworker communities and accounts of how I built rapport with them through participation in community organizing and activism. My *participatory reaction* (Maskens and Blanes 2013) was often driven by farmworkers' concerns and needs, as were the acts of accompaniment and ecosocial solidarities described in chapters 4 and 5. The suturing together of activist and ethnographic methods opened up relationships, reciprocities, and opportunities for study and allowed me to exchange with people in a more mutual and human way (Checker 2005; Hale 2006). It also kept me accountable to the people I worked with, and mindful of how research could contribute to broader and ongoing social, health, and environmental justice goals.

RETHINKING IM/MIGRATION IN THE GLOBAL FOOD SYSTEM

U.S. agriculture has long depended on *immigrants* from other countries who resettled in key food-producing zones, as well as foreign- and U.S.-born *migrants* who travel from place to place, following crops and jobs in farm fields and food processing factories. For example, in the eastern United States, where I hail from, my maternal great-grandmother, an immigrant from Ukraine, and her U.S.-born children (my grandfather among them) migrated from Baltimore to rural southern Pennsylvania over some summers to pick green beans. The extra cash supplemented my great-grandfather's earnings as a worker with the Baltimore and Ohio Railroad.

The farmworkers I met and engaged with came to California seeking work opportunities to sustain their families binationally (Zavella 2011, 2016). Im/migrants themselves hail from diverse places throughout Central America and México. I met men and women from Sinaloa in northern México and Indigenous peoples speaking several different languages from Oaxaca in the south. They entered into some of the most dangerous and lowest paying jobs—jobs that few U.S. citizens to date have expressed interest or willingness to do. Picking strawberries, as I will describe in chapter 2, is one of the harshest jobs.

Many may wonder, if the work is so hard and the conditions so unbearable, why do people come to the United States? Ultimately, for many of the farmworkers I met, it was not always possible to stay in one's home community to try and eke out a living alongside kin and *paisanos*, even if that's what one would prefer. As the physician anthropologist Seth M. Holmes observed while accompanying Indigenous im/migrant Triqui Oaxacan farmworkers, "crossing the border is not a choice . . . but rather a process necessary to survive, to make life *less* risky" (2013, 21, emphasis in the original). It has been estimated that at least half (but likely far more) of the farmworkers living and working in the United States are *undocumented* (Farmworker Justice 2019), a word that encompasses many different paths and experiences. Some farmworkers may have arrived legally through a tourist or temporary worker visa that then expired. Others crossed the Mexican-- U.S. border without a visa, likely with the assistance of a coyote or trafficker, evading official checkpoints and procedures governing international travel and borders, eventually meeting up with kin or friends at their destination. Undocumented victims of human trafficking, sexual or domestic violence, or other crimes who cooperate with law enforcement may be able to secure temporary and then permanent residency through T and U visas, though the paperwork and wait times for these can sometimes be years long. Many of the families I met were composed of members of mixed status: undocumented parents and older children with U.S. citizen children, as well as many other combinations. Some folks who immigrated with authorization may eventually send for family members back home, including children or elderly relatives; this process can take decades because there are strict quotas on the numbers of immigrants that may be accepted into the United States from each country.

From the 1970s through approximately 2010, which marks the span of arrival times for the people I worked with, im/migration from México to the United States rose dramatically following periods of economic instability and restructuring in the 1970s through the 1990s. With the passage of the North American Free Trade Agreement (NAFTA) in 1994, even more im/migrants from México desperately sought opportunities for survival and income. Many of the families I met, including Omar's, used to eke out a living through subsistence farming in their largely rural home communities. They also engaged in seasonal and temporary work in construction and in the underground economy, selling prepared food and other merchandise and services (Zavella 2016). People left their hometowns amid a massive collapse of local agricultural and urban economies and a marked rise in state violence accompanying these transitions. Both are direct consequences of NAFTA (Chollett 2000; Gálvez 2018; Gledhill 1998, 1999; Otero 2011, 2018; Stephen 2007; Zavella 2011, 2016). NAFTA is an economic agreement between Canada, México, and the United States that intended to facilitate trade by reducing tariffs and encourage the growth of industry to create jobs in all three countries.

In reality, however, NAFTA has privileged economic development, investment, and trade for U.S.-based transnational corporations, often to the detriment of workers and rural communities in each country. For example, the agreement contractually forced México to erase many protective and progressive policies designed to protect workers and the environment and took away land rights that had once been guaranteed in the Mexican Constitution. Other multilateral trade agreements have created similar consequences and patterns of im/migration for other countries. Before coming to the United States, many of the farmworkers I met first headed to the extensive and burgeoning corporate farms that are coming to dominate the landscapes of rural Central America and México. There, foreign agribusinesses contract with regional subsidiaries to grow everything from golden raspberries to cocktail tomatoes and coffee beans. Most of this food will not be consumed in the places where it is grown. Instead, it is destined for export to markets in Canada, China, Europe, Japan, and the United States (California Strawberry Commission, n.d.-b). These crops are produced with great quantities of toxic pesticides and fertilizers, and with intensive use of scarce and increasingly privatized and heavily contaminated irrigation water. In turn, this renders agricultural communities on both sides of the border even more vulnerable as residents' bodies, labor, knowledge, and natural resources are devalued, drained, and depreciated.

These patterns of im/migration, settlement, work, displacement, and building and rebuilding community are shared by many rural agricultural workers throughout our global food system (Oxfam America, n.d.; Oxfam International 2018). This makes U.S. mainstream framings of immigration as exceptionally out of control, draining, or dangerous to be quite disingenuous. The massive scale of the global food trade—with its major brand-name labels like Chiquita and Driscoll's; the reassuring comforts of familiar, favorite, and healthy foods at relatively cheap prices at a grocery store near you; and reports of multi-billion-dollar profits for some of the world's largest food and agriculture companies and economies, including the state of California—obscures the multilayered social and ecological vulnerabilities facing farmworkers and their families. The lives, labors, and journeys of Omar and his family, among millions of others, quite literally feed the world, albeit in very uneven ways (Gálvez 2018; Otero 2018). Farmworkers rarely make enough money from these harvesting jobs to pay their rent or buy food, toilet paper, and other basic necessities. They may go into debt from paying a coyote or trafficker, or even a more established friend or family member, just to relocate to work for corporate-owned and -operated farms in order to buy meat and soap and a shared mattress on the floor of a cramped single-family dwelling.

Meanwhile, California agriculture's high-yielding bounty and multi-billion-dollar profitability are celebrated through annual strawberry, pistachio, and other commodity-specific festivals. Agrarian or pastoral romanticism helps market products and creates social identities around familiar foods (Freidberg 2009; Guthman 2004; Pollan 2001). These images certainly tell stories, but they also obscure the broken and breathless bodies of farmworkers—sucked dry, worked to the bone, out of site and out of mind and at the same time in plain sight for those who are paying attention.

This was the case for Oaxacan temporary worker Honesto Silva Ibarra, who died of heatstroke while laboring in the state of Washington's blueberry fields in 2017 for hours on end in hot conditions exacerbated by smoky summer wildfires (Bauer 2019; L. Jones 2017). While the labor contractor was fined for violating break and mealtime laws, the coroner who autopsied Silva Ibarra listed his death as the result of "natural causes" from complications from untreated diabetes.[4] Other farmworkers, like Juan, Lilia, and Milagro (in this chapter) and Aniceta (in chapter 4), sustained multiple and debilitating musculoskeletal and metabolic injuries that made ongoing work in the fields too painful to bear. The prolonged and commingling effects of diabetes and high blood pressure, depression, anxiety, pain, stress, and other chronic conditions also take their tolls (Horton 2016b; Mendenhall 2012) on farmworkers like Berta and Joel (in chapter 2). Mario (in chapter 3) and Gerardo (in chapter 5) each developed neurodegenerative disorders, possibly as a consequence of years of occupational pesticide exposure.

The scope and scale of these health disparities and lost years of work and life cannot be reduced to mere genetics, bad lifestyle choices, or individual behaviors (Holtz et al. 2006). Nor should we write them off as inevitable costs of doing agribusiness and growing cheap food. Food scholars Raj Patel and Jason Moore describe how capitalist economies are sustained through a perpetual "cheapening of nature" (2017, 48). Food workers, including Mexican im/migrant farmworkers, are culturally constructed to be closer to nature, meaning that *they* belong in the fields while others do not. This is evident in California historical documents describing migrant farmworkers as "beasts of the fields" (Street 2004), and in the fact that the first migrant farmworkers in the Americas, slaves imported from Africa, were not considered fully human (Patel and Moore 2017). These myths that continue helping those who profit the most in the food industry justify exploitative labor arrangements and farming styles. Food is transformed from sustenance to money; more value is placed on the potential of crops and farmworkers to generate profits, measured by crop yields over food quality, what sells versus what nourishes, and how fast folks can get the harvest out of the fields and off to market (Patel and Moore 2017, 49, 100). The knowledge, skills, energy, and bodies required to produce and pick food and tend to fields are simultaneously devalued, depleted, and diminished, along with the labor, landscapes, resources, and ecologies from which fruits and vegetables grow (Patel and Moore 2017, 95–96).

Our relationships to each other, and the interconnections between our work, sustenance, and the ecologies we inhabit and share, need not be so extractive and exploitative. To even begin imagining alternatives to our current food and world systems, we must first destabilize and challenge social constructs that render Mexican im/migrant farmworkers as subhuman, undeserving, threatening, nuisances, or inconvenient, and that claim their exploitation is natural and inevitable. Im/migration captures the nuances, fluidity, and diversity of lived and life experiences of the people I worked with more so than the myths about im/migrants presented to us in the news by political figures or others with a stake in keeping the food system just as it is. It is unreasonable, given the histories and relationships I've described in brief, to keep treating México or Mexicans as politically and socially separate from our existence in the United States. We must acknowledge our connections and interdependences with rural peoples and places, or the "countrysides" (Patel and Moore 2017, 103), where food and im/migrant farmworkers come from, as mutually constituting one another. Tackling the myths that ground industrial agricultural production models and perpetuate social disconnection and alienation are helpful starting points for activism.

ANTI-IMMIGRANT RACISM: MYTHS OF MOOCHERS, MALINGERERS, AND MALIGNANCIES

Upon arriving in California in 2010, one of my first field notes described a conversation I had in a tourist town a few hours away from Watsonville. A group of white men, enjoying their vacations, upon hearing about my research goals, felt that the im/migrant Mexicans who worked at the hotels and restaurants in the area were lucky to live in California, a "paradise." They did not explicitly hate or resent im/migrants, but they also could not imagine why someone living or working in the United States would have cause to grieve or complain. After all, hadn't César Chávez taken care of farmworkers' plights almost fifty years ago?[5]

Others expressed more overt hostility and racism toward im/migrant farmworkers and their families. The *Watsonville Patch*, an online news and community forum, featured an announcement about a used clothing and household item drive organized by the Center for Farmworker Families, which I worked with regularly throughout my fieldwork. The center's free flea markets, or *tianguis gratis*, were largely accomplished through volunteer labor and from donations of gently used and new items, mainly from wealthier supporters in the San Francisco Bay Area. The *tianguis* allowed farmworker families to "shop" for clothing, household, school, and other supplies during the off-season of November through April when funds dwindled. At Christmastime, the *tianguis* became especially lively and sometimes ended with a piñata party. I spent a lot of time collecting and sorting donations and organizing and promoting the events among farmworkers through handmade flyers and by word of mouth. The back of my car regularly filled with bags of clothing and other odds and ends. Requests for home furnishings came often, given how unstable and unaffordable housing is on California's Central Coast. For example, farmworking sisters Yesenia and Yolanda moved a total of six times over the course of three years—from a walk-in closet at a coworker's house, to a women's shelter, to a shared room, to a converted tool shed in someone's backyard, to a recreational vehicle, and finally to a studio apartment. A California Institute for Rural Studies report suggests that this is more the norm than the exception (CIRS 2018).[6] The center often helped farmworkers find mattresses and bedding. Folks at the free flea market gatherings sometimes became my research participants and friends.

I wrote a press release soliciting donations for a *tianguis*, published in good faith by the then editor of the *Watsonville Patch*. It elicited responses in the comment feed such as, "Why don't *they* [event organizers] spend time helping *their own*?" ("Their own" implied legal citizens of the United States.) In the words of one commenter, "Mexico is an awful place for unskilled laborers. They are treated like slaves. All of them are staying here [in Watsonville]. If they can't find a job— no problem—they just figure out a way to go on public assistance. Have the female get preggy and have the baby in an American hospital! BINGO! Lottery time. Free shelter, free food, free medical care, plus free spending money compliments of the US and CA taxpayers!" This response reproduces extremely racist tropes devaluing im/migrants as unskilled workers and Mexicans as inherently bad and parasitic people. It ignores the roles and behaviors of U.S. foreign policies and corporations, described in brief above, in producing and exacerbating poverty in México and elsewhere in Latin America. The comment poster lambasts im/migrants as the undeserving recipients of free goods, services, support, and caring emotions. He or she insists that im/migrants' existence in the United States places undue burdens on more deserving citizens who struggle to find work and survive amid ongoing economic crises (Chávez 2013).

Throughout my field research, it initially surprised me how many people assumed that *all* farmworkers had plentiful no-cost benefits such as free housing, health care, childcare, and food assistance. In reality, people's situations were much more complicated and their access to support far more limited (see chapter 4). What I observed in farmworker households and gleaned from listening to their stories contrasted sharply with these assumptions that im/migrants have access to, even aggressively seek out, these nonexistent "free lunch" buffets of health care, housing, and other safety net services when they cross the border. Such supports are not even easy to access for U.S. citizens who need them—let alone im/migrant farmworkers. Access requires hours of paperwork, eligibility screenings, interviews and assessments, waiting in line (Gupta 2012), and figuring out, logistically, how to get from point A to point B, sometimes with children in tow.

In another instance, at a Santa Cruz City Council meeting I attended in 2011, a concerned citizen blamed undocumented immigrants for depleting California's redwood forests, which was the dubious basis of his vehement argument in favor of preserving the Secure Communities program. After 9/11, the Secure Communities Act gave ICE permission to work directly with local law enforcement in order to find, detain, and deport immigrants convicted of crimes (Kline 2019). After people were processed and fingerprinted at city and county jails, their data would be entered into federal databases, including those cross-listed with ICE. Individuals who had previously tried to enter the United States and had been detained at the border would come up in the searches, triggering ICE to intervene even if they had not been convicted of anything. Detainees would be transferred to federal detention centers, where they awaited deportation. Secure Communities granted ICE agents the power to screen and hold people "solely based on immigration status" (ACLU Northern California, n.d.). Being caught in this legal limbo caused significant problems for families of mixed immigration status, even for something as benign as a minor traffic violation, or even when accused individuals had committed no offenses at all. To counter these legal violations and ambiguities, the California TRUST Act went into effect in 2014. It attempted to reduce cooperation between local law enforcement and ICE, mitigate harms caused by family separations, and restore community confidence in law enforcement. Loopholes granted individual counties and cities discretion as to which crimes would be prioritized for extended detention in local facilities or deportation via ICE, creating uneven protections for im/migrant communities throughout California (Ulloa 2017).[7]

The commenter at the Santa Cruz City Council meeting thought the Secure Communities Act would help protect California's natural resources—or at least the redwood trees. Similar illogic reverberated in a 2015 advertisement launched by the anti-immigrant group Californians for Population Stabilization. On YouTube and on national television, a young white boy asks seemingly innocent questions about the relationships between California's rising population, immigration, and diminished water supplies. In the end the ad directly accuses immigrants of causing the statewide water shortage, insinuating that overconsumption by communities was responsible for depleting groundwater. In reality, California's agricultural industry uses the largest quantities of water (Gong 2015). In the case of California's famous redwood forests, lumber and paper supply companies; construction; and the consequences of climate change, such as drought, disease, pests, and increasingly large and long-lasting forest fires—not im/migrants—play the most significant roles in forest destruction.[8]

Discourses about im/migrants as thieves, frauds, resource drains, environmental polluters, and public health hazards abound and have deep roots (Cagle 2019; Chang 2000; Huang 2008; Kulish and McIntire 2019; Molina 2006; Schelhas 2002; Park and Pellow 2011). They deserve our close scrutiny and skepticism, especially as they increasingly reverberate globally. Other countries are experiencing rises in nativist movements and the shaming of poor people, from Brazil to Finland. Immigrants and their descendants who internalize these shaming discourses may refuse assistance or reproduce harmful narratives of individual responsibility and victim blaming in their own lives and communities (Quesada 2011). Latinas in the United States face additional scrutiny through social and political preoccupation with their fertility, as evidenced in the online commenter's suggestion that getting pregnant is a reasonable and viable way for im/migrant women to access more resources. Meanwhile, data suggests that Latinx families are not significantly larger than those of whites or U.S. citizens (Chávez 2004; Gálvez 2011; E. R. Gutiérrez 2008).

Inadvertently, anti-immigrant narratives also facilitate the gutting of programs and supports that benefit large numbers of people, including millions of American citizens. They also embolden racialized animosities, violence, and threats, as the summer 2019 mass shootings in Gilroy, California (not far from the Pájaro Valley) and El Paso, Texas, demonstrate (Sanchez 2019). At one point, an especially vocal online commentator showed up at a *tianguis* event. He informed one of the on-site organizers that he had a knife and was not afraid to use it if any of the "illegals" present threatened his safety.[9] Most of the folks who came to the *tianguis* events were women with small children (husbands and partners tended to hang back in their cars, on the curb, or at home) and elderly men. Alerting a Watsonville City Council representative and the police about this individual resulted in no significant consequences.

Another pervasive misconception about im/migrant farmworkers, especially undocumented people, is that they are frauds (Horton 2016a, 2016b) who use fake—or worse, stolen—Social Security numbers to gain access to jobs, benefits, or resources like food and housing assistance and health care. Irrespective of citizenship or residency status, im/migrants overall consume a much smaller proportion of state and federal aid, health care services, and natural resources than U.S. citizens, in large part because their access to it is unequal and inequitable (Berk et al. 2000; Cha 2019; Chávez 2003; Mulligan and Castañeda 2018) and limited by fear, shame, geographic distance, language barriers, and income constraints (Berk et al. 2000; Carney 2015; Pitkin Derose, Scarce, and Lurie 2007; Quesada 2011).

Medical anthropologist Sarah Horton (2016a, 2016b) has observed how, in reality, some undocumented farmworker families engage in the practice of identity loan, wherein the Social Security card of one documented family member is shared among others within a kin group as a way of ensuring access to work and the pooling of resources.[10] This practice differs significantly from accusations of theft and fraud that are being used to make policies that end up curtailing citizen and noncitizen access to desperately needed supports. Identity loan is not significantly different from other methods impoverished communities have used to pool and share resources in other eras and contexts (Stack 1970). The fraudulent activities of employers and contractors, who knowingly hire and rely on undocumented im/migrant workers, receives significantly less scrutiny and moral shaming (Horton 2016a, 2016b). Contractors and growers may be subject to relatively small fines for failing to comply with laws, such as pesticide or other occupational health, safety, and wage policies. The farm where guest worker Honesto Silva Ibarra worked before dying from heat exposure received a nearly $150,000 fine from the Washington state Department of Labor and Industries for violating break, mealtime, wage and hour laws, but not for killing Silva Ibarra (Associated Press 2018). The threat of fines is not always enough to prevent infractions that jeopardize farmworkers' health and lives. The death of Indigenous farmworker Maria Isabel Jiménez in 2008 from heat exhaustion has been followed by year after year of subsequent farmworker deaths due to heat, despite efforts to strengthen farm workplace and heat safety regulations (Horton 2016b).[11]

In these and other ways, undocumented im/migrants are excluded not only from their legal rights as workers but also an array of social and health assistance programs that they directly (through paychecks) and indirectly (through cheap labor that lowers our costs of living) contribute to (Zavella 2016). Paying into the safety net but not being protected by it is yet another way that racialized im/migrant labor (Holt-Gimenez 2017a, 2017b; Polidor 2017) subsidizes industrial agriculture's profits and the health and overall welfare of the general population, as the stories of Juan, Lilia, and Milagro illustrate.

JUAN: THE LIFE COURSE OF INJURIES

I met Juan at a workers' compensation legal clinic at the Watsonville Law Center, where lawyers volunteered to review injured farmworkers' cases. Pro bono lawyers offered suggestions, not formal legal advice, on what actions, if any, farmworkers could pursue. I had the opportunity to sit in on some of these meetings at the clinic. Interpreters relayed information between the farmworkers and the lawyers, who occupied a separate conference room. Farmworkers entered into smaller offices where they could have some privacy with their interpreters.

As the workers stopped by, I let them know about my interests and asked them if it was all right for me to document their stories. Most did not mind. Even when they knew that they had exhausted all of their options, they wanted to be heard.

Juan slowly, cautiously, lowered himself into a cushioned office chair. He told me that in 1990, while working for a unionized lettuce harvesting company that would later merge with a larger nonunion firm, Juan had felt a painful twinge in his lower back. His work involved deftly slicing the heads of lettuce at their stem, all the while maintaining a stooped position. From there the lettuce heads were tossed up to another crew member perched on a motorized conveyer belt. They were speedily wrapped and packed into waxed cardboard boxes, trucked to the cooler, and then shipped off to a wholesaler or retailer.

In his mid-twenties at the time, Juan filed for and received workers' compensation coverage from his employer's insurer. The insurance company approved a back surgery proposed by his doctor to repair two herniated discs. Even after surgery, and disability leave and unemployment from 1990 to 1993, Juan's pain continued to worsen. Nevertheless, he returned to work for the same crew for ten more years, from 1993 to 2003.

In 2003, now at age thirty-five, he reinjured himself. He could hardly walk, let alone efficiently heave heads and crates of lettuce at a quick pace, and thus could no longer bear to work.

For almost another ten years Juan's case was in the courts, which eventually determined that Juan's employer at the newly merged giant lettuce firm was not responsible for covering his current issues. The employer argued that the new injury resulted from an older injury sustained at his former job with the then smaller firm. The merger had voided the new firm's legal obligations to attend to workers' so-called preexisting work-related injuries. The original insurance company also no longer existed. The paperwork Juan and his doctors signed in 1993 showed that they had closed his case in exchange for a onetime settlement. It would now be impossible for him to receive follow-up or continuance care and coverage through workers' compensation. Many farmworkers are tempted or coerced to settle in these cases because it is so unbearable, physically and emotionally, to be left waiting for courts to make rulings (Saxton 2013).

In the meantime, as a legal resident Juan received monthly checks from Social Security Disability Insurance (SSDI) and additional health care support from Medi-Cal—the state-sponsored health insurance program for low-income individuals and families in California. Undocumented farmworkers, even though they pay into Social Security with each paycheck, are ineligible for SSDI and unemployment benefits (Horton 2016b, Zavella 2016). The one treatment that brought him some relief, physical therapy in the pool at the YMCA, became cost prohibitive. Medi-Cal and SSDI would only cover so many visits, and Juan had reached his reimbursable limit. Paying out of pocket for therapy created financial strains for his family of five, which included three young daughters.

The only expense the original workers' compensation insurer continued to cover were prescription muscle relaxers and painkillers. Juan did not like taking these; they made him feel dizzy and dysfunctional and limited his ability to spend time with and take care of his children. He reflected on the effects of different treatments on his pain levels and sense of overall well-being:

> La terapia me relaja y me da animas a seguir con mi vida . . . las drogas me afectan a mi mente, es como mi mente está durmiendo, pero mi cuerpo funciona. La terapia no me cura, pero me da fuerza. No entiendo por qué la seguranza no cubre algunas cosas. ¿Por qué la autorización de mi doctor tiene valor en estos casos, y en otros no?

Physical therapy relaxes me and encourages me to move forward in life . . . the drugs affect my mind; it's like my mind is asleep but my body is functioning. Physical therapy doesn't cure me, but it gives me strength. I don't understand why the insurance does not cover some things. Why does my doctor's authorization have legitimacy in some cases, and not in others?

Juan clearly understood that he would never fully recover to his preinjury state, but he still wanted to be able feel well enough to be a good father and husband. Physical therapy gave him a reason to *seguir adelante* (keep going). The exercises provided both physical and emotional relief. The feelings of helplessness and inadequacy as a man due to his inability to work (Horton 2016b) made him depressed.

An MRI in 2008 showed several more herniated discs. Doctors recommended additional surgery to repair these in addition to errors from the first surgery. Juan felt wary about this. He knew surgery did not guarantee him relief and could very well leave him in far worse shape: permanently disabled by pain and physical trauma or, worse, paralyzed. This is a dilemma faced by many work-injured and disabled farmworkers: they must choose between a life with chronic yet knowable (and in the best circumstances, manageable) pain or the uncertainties of prescribed surgeries, drugs, and other procedures.

When Juan walked into the Watsonville Law Center, he had a feeling that there was nothing anyone could do for him legally. Nevertheless, getting closure helped him strategize his health care decisions. Social security disability payments helped ease some financial and medical burdens but could not sustain his family.

Juan anchored his arm to the chair to support his weight. He rose to shake our hands, a demonstration of respect, but a move that was also necessary to provide relief. His voice trembled with pain as he explained that he would be more comfortable standing and waiting in the office. His foot had gone numb and the pain was shooting up his spine.

The lawyer's response was as he expected: "Good luck," she said. It was very unlikely that the employer's insurance would help pay for physical therapy. When Juan had signed the settlement papers in 1993, that relieved the company of any financial responsibility. No workers' compensation lawyers would take Juan's case because they only get paid when permanent disability payments are awarded. Even though the original 1993 paperwork stated, "There may be need for future medical treatment," the lawyer felt that Juan was lucky that the insurance company continued to pay for his prescription medicines.

When Juan received this feedback, he did not seem too disappointed. It was a response he had anticipated. "I had these doubts, for many years, is it right, or was there something else I could do? The company never asks us workers how much to take out of the check for insurance. I just wanted to know." We shook hands with Juan, exchanging sympathetic looks. We all knew it wasn't right, but legally, there was no recourse. The legal clinic receives many cases like Juan's every month. All it can do is educate farmworkers about their workers' compensation, health care, and other legal rights, however limited.

Ultimately, the argument that injured farmworkers game the system to earn income, get free health care, and avoid working is dubious. Farmworkers are paid so little; any compensation for lost wages amounts to a very small fraction of already inadequate salaries, and there are many attempts by employers and insurers to abdicate and avoid responsibility for medical care. Workers' compensation cases also take a really long time to process and involve intrusive investigations that lead many farmworkers to give up, settle, and go back to México disabled (Unterberger 2018) or to return to work even before they have adequately healed or recovered.[12] This has consequences not only for individual injured workers but for entire families and binational community networks, as the stories of Lilia and Milagro illustrate.

Interfamilial, Intergenerational, and Intersectional Injuries

I met Lilia, a thirty-year-old farmworker from Michoacán, México, at a sewing circle at a Catholic church. Hermana Blanca, one of the resident nuns, saw the need for a space where farmworker women could socialize and be creative together over the long, lean, and isolating winter months. Women brought their knitting, crochet, and embroidery projects with them, hosted monthly potlucks, and threw baby showers for one another.

Not everyone in the circle warmed to me. Hermana Blanca explained that it was hard for many of the women to trust strangers after enduring rape, domestic violence, human trafficking, abuse at work, and general exploitation. Some of the Indigenous Oaxacan women, who were also some of the most talented artisans, spoke little to no Spanish and faced racialized animosities from other group members. Still, I brought embroidery cloth and cotton thread to share. I attempted to relearn knitting and failed, but I had better success making hand-knotted friendship bracelets.

Hermana Blanca sometimes invited guest speakers to come share information about programs serving farmworkers. At one point, I gave a presentation on pesticides and soil fumigants. This got Lilia's attention.

When we met in 2011 Lilia had been working for the past two years at a vegetable packing plant and, prior to that, several years in the strawberry fields. After bonding over her chunky-cheeked newborn son and the delicate hand-crocheted sweaters she adorned him in, she invited me over to her brother's house. In the off-season, Lilia had been watching over her three nephews, ages one to ten years at the time. *La migra* (slang for ICE) had recently detained her *cuñada* (sister-in-law) at a facility three hours away from Watsonville. Lilia's brother—employed as a foreman at a large strawberry growing firm—spent most of his time trying to stop his wife's deportation.

I sensed my presence provided a welcome distraction and a chance to vent. Lilia's face showed signs of worry and distress as she balanced caring for her own two boys in addition to three traumatized nephews. The youngest baby cried all the time. The eldest had grown more and more withdrawn.

La migra had hurt their family before. During a workplace immigration raid in the strawberry fields, Lilia's husband, David, ran toward a nearby storage shed and tumbled into an irrigation ditch, injuring his knee. He evaded *la migra*, but his knee pain persisted. It became difficult to walk and perform the repetitive stooping and bending motions in the strawberry fields. He didn't have insurance and did not want to draw more attention to his employer, so he decided not to seek treatment.

David needed to keep his job, but he could no longer physically endure full-time hours in the strawberry fields. He found a part-time job at a composting plant, which took some of the pressure off his body since the movements of the work were more varied; there he received an hourly rate. Still, Lilia sometimes literally had to unbend her husband's stiff body so that he could get out of the car, a chair, or their bed.

This is one reason why Lilia sought work at a packing plant. Although she only worked seven months out of the year, the job provided her with basic health care benefits through an on-site clinic ran by a subcontracted provider. Eventually Lilia's knees started to ache too from the constant standing on hard concrete floors in the cold packing plant. Her shoulder also hurt from the repetitive motions of sorting delicate baby lettuces on a fast-paced conveyor belt. She attributed her regular migraines, which flared up during the work season, to regular exposure to chlorine bleach. This is why my talk about pesticides and health piqued her interest. Bleach is used in a diluted solution to remove dirt, debris, and potentially deadly pathogens like *E. coli* from leafy greens so that they are market ready. The fumes wafted through the air in the plant, commingling with the vapors from ammonia-based solvents and industrial-strength cleaners used by the night crew to sterilize factory equipment.

Unlike Lilia and her workmates, the cleaning crew had a union. They recently had gone on strike, protesting their routine exposure to noxious chemical fumes. The women did not have a secure way to address their own toxic workplace exposure without risking losing their jobs. Instead they coped with their frustrations by sneaking small amounts of lettuce into their apron folds. At least their families could enjoy *ensaladas* (salads) at home.

Aside from taking disability leave for the birth of her youngest son, Lilia never requested time off work or filed for workers' compensation for her knee or shoulder pain or her chronic headaches. Doctors at the factory clinic just gave her prescriptions for each complaint. Whenever she had extra cash, she visited a *sobadora*, a Mexican traditional healer who integrates massage and musculoskeletal manipulation to help relieve pain and provide spiritual comfort. *Sobadores* charge less than conventional chiropractors, massage therapists, and clinics: around twenty-five to thirty dollars per session, as can be seen advertised on billboards and brochures left strategically at laundromats and *panaderias* (Mexican bakeries) frequented by farmworkers.

Years before, when Lilia worked in the strawberry fields, she suffered a miscarriage. She and other farmworker women whom I interviewed referenced miscarried or stillborn children when composing kinship charts with me, valuing their places on the family tree alongside their living siblings. Neither these children's lives nor those of the slowly deteriorating farmworkers and packing plant workers are accounted for in official agricultural injury and death tolls. This prevents necessary proactive measures from being taken and allows companies to continue profiting while denying responsibility for the problems business and workplace practices create and perpetuate. As Lisa Cullen acknowledges, "No other public health issue . . . is treated as cavalierly as occupational health and safety" (2002, 54), especially for low-wage workers. Lilia's status as an undocumented im/migrant farm and food worker—with limited access to health care; a precarious position on the packing plant assembly line; multiple caregiving responsibilities (mother, wife, aunt, sister, friend); and severe social and physical stress and pain—compound and exacerbate one another.

"Se Renta un Garaje" (Garage for Rent): The Myth of American Dreams

The myths I've described so far distract us from the real roots of the widening gaps between rich and poor in the United States (Greenbaum 2015) and globally, which are linked to increased and unfettered control over wealth and natural resources by a select, relatively small group of elites and their corporations. The relatively low costs of food for nonfarmworking consumers is subsidized by the underpaid work of im/migrant farmworkers and their long-standing legal and social exclusion from rights and protections at work and elsewhere. Anthropologists Nick Copeland and Christine Labuski (2013) argue that U.S. taxpayers subsidize Walmart's multibillion-dollar profitability. Walmart's underpaid and unbenefited employees must rely on the state-subsidized federal health care program Medicaid and on federal food assistance and housing support just to get by, providing a hidden subsidy to Walmart's profits. Meanwhile, an endless supply of cheap goods palliates the suffering that comes with economic hardship in the general population but does little in the end to address systemic inequalities that cause poverty. Ultimately, the company's promise of jobs in communities where opportunities are few and far between has not been sustainable, as stores are shuttering more often than opening.

Our cultural and political aversion toward impoverished people, im/migrant and otherwise, who are perceived to be freeloaders, moochers, malingerers, and drains on the system, is not well grounded in the lived realities of low-wage workers, including im/migrant farmworkers. None of the farmworkers I met during my field work received free housing. Very few had access to subsidized dwellings, such as in the Section 8 Housing Choice Voucher Program, or camps like the one where Berta and Joel lived (see chapter 2) that were designated for farmworkers and partially supported and managed by federal, state, county, and nonprofit funds. The overwhelming majority of the farmworkers I met lived in decrepit single-family homes or apartments shared with multiple families and individuals, or in outbuildings, garages, old barns, shacks handmade with cardboard or plywood (*casas de cartón*), and aluminum-sided sheds meant for tools, sometimes lined with makeshift insulation of old carpeting or newspapers. Rents ranged from five hundred dollars to over sixteen hundred dollars a month for a single bedroom (or garage or shed) and did not change significantly with the quality of the dwelling or the number of people renting.

Zavella (2011) describes the coexistence of low-wage and underpaid jobs and one of the highest costs of living in the United States on California's Central Coast as symptoms of the uneven processes of development and globalization. These patterns of intense and rapid development catering toward those with concentrated wealth in the state have been driving housing prices and rents up. This is squeezing farmworkers and other residents, including teachers and health care workers, out of safe and affordable places to live. Processes of uneven development and globalization are also what lead many im/migrants to come to the Central Coast in search of work in the first place. Neither Juan's, Lilia's, nor Milagro's families enjoyed any of the mythical spoils of the American dream, which proclaims that people who work hard will not only survive but thrive. In the long run, hard work in agriculture amid injurious, toxic, and impoverishing conditions, with little recourse and few resources for support and healing, does more harm than good.

I conducted an interview with Milagro, a work-injured farmworker whom I also met through the sewing circle. Milagro worked in the strawberry fields around Watsonville from May through November. In the winter months, she and her husband would migrate to the strawberry fields in Oxnard, farther south on the California coast, about an hour north of Los Angeles.

In their rural *rancho* (small village) in Michoacán, México, Milagro's parents could not afford to send her or her twelve other siblings to school beyond the sixth grade. From age twelve onward, Milagro worked with her mother, picking tomatoes on corporate farms for cash and taking care of the family's farm animals. Her father, also a farmworker, migrated back and forth between California and México, working in the very vineyards and strawberry fields where Milagro would one day labor. Complications from a stroke permanently disabled him at age fifty-five, by which time he had already returned to México. Milagro's mother suffered from advanced diabetes, but still took care of her many grandchildren who remained in México, unable to cross the border and be reunited with their parents. This included Milagro's two eldest daughters.

At seventeen Milagro followed her older brothers to California. Maybe, together, she and her brothers could save up enough to build their family a better house. They migrated back and forth over the years. Milagro started a family with her ex-partner; when it came time to return to California for work, she left her daughters in the care of her mother. At first, she thought, she could send for her daughters once she felt established and secure in her work. She said with a sigh, "Esos eran mis ilusiones de venirme" (Those were my false hopes for coming here). Going back and forth across the Mexican-U.S. border became too difficult, and eventually, years had passed since she last saw her eldest girls. Since then, she had another daughter with her new partner in the U.S.

Milagro lit up as she bragged about how fast she once picked strawberries, bringing in checks of up to eight or nine hundred dollars every fifteen days: an impressive pace, but not one that she nor many others could sustain in the long term. This is another paradox of farmwork: many workers take pride in their ability to work hard and push their bodies to their limits. Some employers and community members in farmworker communities perceive work ethic and physical ability in agriculture as an innate biological capability possessed by Mexicans (Holmes 2007; Horton 2016b); it becomes a sense of community identity. The sadly comical inability of U.S. comedian Stephen Colbert to keep up with farm and packing plant workers in upstate New York in a series he produced on immigration comes to mind (Colbert 2010). The clip aims to foster greater empathy for farmworkers, but also inadvertently normalizes Mexicans as naturally suited for agricultural jobs that seemingly no one else wants to do. Similarly, the United Farm Workers' (UFW's) Take Our Jobs campaign, featured in the Colbert episode, argued for a path to legal status for agricultural workers and further rights and protections irrespective of their citizenship status (United Farm Workers 2010). It did little, however, to challenge the origins or continuance of racial divisions of labor in agriculture because it reiterated that no one else could or would work in the fields other than Mexican im/migrants.

The piece rate system of pay, in which workers are paid by the bucket, flat, or crate picked or plant sewn, is typical for harvest work, whereas other kinds of farmwork, like weeding, pruning, thinning, or clearing fields are paid by the hour. While piece rates are intended to reward workers for their efforts, they also can make workers more vulnerable to work-related illnesses and injuries and chronic stress and strain. Farmworkers feel a great deal of pressure to go hard and fast in the fields to prove their value and to not be perceived by peers, contractors, or employers as lazy or deficient (Holmes 2013, Horton 2016b). They must also keep up the pace in order to earn enough money to support their families on both sides of the border. One's financial success as an im/migrant worker becomes a significant marker of social worth and dignity in home communities

(Mahler 1995). The inverse is also true, where failure to earn a living in the United States or México is marked by stigma and shame (Unterberger 2018).

In 2009, when she was twenty-seven years old, Milagro felt a sudden, sharp, and paralyzing pain while harvesting strawberries in a stooped position. At first, doctors affiliated with the workers' compensation insurance company that her employer contracted cared for her for a month and eventually diagnosed her with muscle spasms. A different physician who examined Milagro on behalf of a lawyer concluded that she had three herniated discs and that her back was permanently injured.

It took a lot of energy, both physical and emotional, for Milagro to manage her pain and get used to doing everything at home at a more slow and labored pace. Similar to Juan, she also experienced sharp pains, along with numbness in her legs and feet. Limiting her diagnosis to muscle spasms of the back did not take into consideration how the spine supports other parts of the body necessary for daily life and functioning. Over time, one injury can evolve into several coexacerbating problems. As she explained, "Todavía me lo está afectando, pero como le digo me han puesto dos inyecciones en la espalda y eso es lo que me ha ayudado. . . . Porque yo no podía caminar nada. Andaba bien jorobada como viejita." (It's still affecting [my back], but I tell you they have given me two injections in the spine, that's what's helped me. . . . Because I couldn't walk at all. I walked all hunched over, like a little old lady.)

Even with intense chronic pain and mobility issues, Milagro scrambled to find ways to supplement her family's income. She could no longer work in the strawberry fields. She ran an informal day-care center in the small converted garage where she, her husband, and her four-year-old daughter lived, watching over the children of her former coworkers. During my visits the kids mostly hung out in front of the television watching cartoons and movies in English.

Limping with the support of a cane as she prepared lunches and snacks for the children under her care, she told me that she wasn't the only one in her family to lose her job. When she started working with a lawyer to help her file a formal workers' compensation case, Milagro's husband, Erasto, stopped receiving calls to come in to work. This kind of retaliation is illegal, but not uncommon, as employers can justify letting people go for economic reasons without penalty. Ultimately, this put more strain and stress on their family. The house Milagro wanted to build for her parents, daughters, and other kin in México would be postponed indefinitely.

During a 2016 trip to San Quintín, Baja California, where other groups of farmworkers emigrated to labor in in the strawberry, tomato, cucumber, and cut flower fields, the remains of incomplete and abandoned small cinder block homes dotted the dirt roads of the *colonias*—worker community settlements that sprang up around towns to accommodate the influx of emigrants. The haunting skeletal forms of these incomplete homes provided more evidence that Mexican im/migrants' modest desires to live with dignity and provide for their families could not be so easily fulfilled, especially if one suffered disability or illness as a consequence of injurious, toxic, and harmful working and living conditions.

Eventually Erasto found work with another contractor. I sometimes wondered what would happen if he and countless other farmworkers in similar situations were to suffer a debilitating injury—or worse, a work-related death. Even when families, *paisanos*, or communities pooled or shared resources like housing—what's sometimes referred to in American cultural discourses as cutting back or living within one's means—they still faced precarities that posed hazards to their health and well-being.

I thought to myself about how close in age I was to the farmworkers who talked with me. Their life opportunities and potentials had been severely curtailed by work-related injuries and illnesses and conditions of poverty on both sides of the border. The places where they lived added insult to injury and ultimately contributed yet more layers to already trying health issues.

BUSTING MYTHS ABOUT IM/MIGRANT FARMWORKERS

From coffee to cut flowers, im/migrant farmworkers all over the world are recruited to harvest food and other specialty crops. Their availability and desperation hinge on their displacement from their home communities by drought, hunger, unemployment, conflicts, political violence, unfair trade agreements, and other unnatural disasters (Fjord and Manderson 2009). They face harsh and violent treatment, harassment, and discrimination at work and in the places where they settle to try to live, even if impermanently.

The broader political context of Mexican migration shows that, in reality, Mexican lives, livelihoods, lands, and resources have been systematically upended, stolen, and jeopardized. Anti-immigrant mythologies frame im/migrants, including farmworkers, as drains on the U.S. economy and safety net systems. Meanwhile, housing, food, and health care are becoming increasingly cost prohibitive for many groups, and especially low-wage and underemployed folks like farmworkers. Many im/migrant workers do pay income taxes and pay into Medicare and Social Security even though they are legally prohibited from collecting these benefits (Horton 2016b; Lantigua 2011).

The false romanticization of farmwork as a simple "lifestyle," as unskilled and mindless work, or the idea that conditions in the fields have improved compared to previous eras are also based in myths and misconceptions. Activist labor historian and former UFW organizer Frank Bardacke (2011) worked with lettuce and celery harvesting crews in the 1970s. He and his highly skilled *compañeros* could earn upwards of twenty dollars an hour (in 1970s dollars) with full health benefits under a union contract. Some of these workers went on to buy homes in and around Watsonville or started small businesses, creating jobs for other family members. Bardacke describes how the demise of the farmworkers' union and the rise of free trade agreements like NAFTA in the early 1980s led to a sharp decrease in wages and protections that persists into the present day. Proposals to raise minimum wages (Martin 2011) will only help so much if the costs of living continue to mount. Resistance to wage increases within the agricultural industry is fierce, despite exponential and sustained growth in corporate profits (Ramey 2019).[13]

Myth busting allows us to critically resist how undocumented im/migrants are framed as a net drain or plague on our society. Too often, even when im/migrants are celebrated in political discourses, they are reduced to their economic value as generators of jobs and profits in vital industries like agriculture. In reality, farmworkers' labors, everyday struggles and hardships, and lost lives and livelihoods in México and the United States subsidize the costs of food for growers and consumers. The widespread and long-standing poverty endured by agricultural workers is not a natural condition. It is rooted in the history of U.S. agribusiness: an economic and ecological system of extraction and monoculture, dependent on the unpaid and underpaid labor and farming and botanical and biological knowledge of Indigenous peoples and Black slaves. Their lives, labors, and wisdom generated wealth for colonial land and business owners (Patel and Moore 2017). Today thousands of farmworkers toil for poverty wages, with few to no protections, continuing the cycle of profit streams for companies as wide ranging as Chiquita, Driscoll's, and Starbucks. The promises of economic growth, prosperity, and progress for people living in the countries and communities where these businesses source their products remain elusive for most ordinary folks.

The tens of billions of dollars generated by California agriculture (MCAC 2017; Santa Cruz County 2017) could very well be distributed in more equitable ways to the benefit of farmworkers, growers and farmers, and agricultural communities. The globalized hyperhierarchical and harmful systems of agricultural production are not normal or inevitable. They shape farmworker health and well-being and foster far-flung social, health, and ecological consequences. A discussion of the history and political ecology of industrial strawberry agriculture and the laws that govern and at times inadvertently enable abuse, suffering, and toxic exposure of im/migrant farmworkers are the foci of chapters 2 and 3.

2 · STRAWBERRIES
An (Un)natural History

"Enjoy 8-a-Day" urges the California Strawberry Commission, the primary trade group for the industry (California Strawberry Commission, n.d.-c).[1] In contrast, early twentieth-century plant scientist S. W. Fletcher remarked that strawberries did not have much nutritional value (1917, 181–187). After all, they are mostly water, just like many other living things. By the twenty-first century, however, the strawberry industry—its trade groups, vertically-integrated grower-shipper companies, and network of farmers and growers—had transformed strawberries and other berries from occasional treats into "superfruits" to be included in our daily diets. Visit almost any major supermarket, and fresh strawberries and other fresh fruits and vegetables—uniform in size, shape, and color—are available for purchase. The ubiquity of strawberries and other berries is visible in a variety of forms: frozen, processed into jams, pureed with apples in apple sauce, blended into yogurt and ice cream, freeze-dried and added to breakfast cereals and granola bars, baked into breads and cakes, tossed into salads, emulsified into dressings and sauces, and adulterated with artificial flavorings and colors to be used as fillings in things like toaster pastries, candies, and other snacks largely marketed toward children.

Once only a seasonal delicacy, berries are today shipped far from their point of origin and consumed all over the world, sometimes as a status symbol. At the 2008 Summer Olympics in Beijing, the Chinese government lifted an agricultural food trade ban with the United States, at least for the duration of the games. Athletes and sports enthusiasts specifically requested strawberries to be one of the official snacks of the games. California's mass quantities of strawberries sealed the deal, with much fanfare from growers and the California Strawberry Commission (Kan-Rice 2008; Raine 2008; Steinhauer 2008). Could California strawberries facilitate international diplomacy? Current and ongoing international trade disputes between China and the United States aside, berries and the capital they garner move far more freely across borders than the im/migrants who harvest them.[2]

In addition to their role in international affairs and sweetening trade deals, the potential sexual potency of strawberries is invoked in a readily Googled photograph of Angelina Jolie, suggestively biting into a big red berry with her famously plump and painted lips. Around Valentine's Day in the United States, the romantic and erotic potential of strawberries goes on sale; they are coated in chocolate and dipped in whipped cream. In contrast, multiple decades' worth of iterations of the children's cartoon *Strawberry Shortcake* invoke images of sweetness and innocence. The program's characters are anthropomorphized into berries, sweets, and other fruits who bide their time gardening, baking treats, and warding off evildoers. In medieval manuscripts and Renaissance art pieces, including *The Garden of Earthly Delights* by Dutch painter Hieronymus Bosch, delicately painted strawberries symbolized rebirth and righteousness, sin and danger, or sexual promiscuity and the temporality of pleasure in the life course, depending on which art historian you ask (Bowen 2011; Bruggmann 1966; Gibson 2003; Stewart 2019).

Mixed, confounding, and ever evolving metaphors and symbolism aside, today strawberries are a hot (and carefully chilled) commodity, a favorite flavor, a gesture of affections, a nostalgic memory, a pull for im/migrant workers, a job opportunity, and a source of regional pride and identity. Each year in early August, Watsonville, California, hosts the Watsonville Strawberry Festival. People convene downtown to taste strawberry-forward delicacies and raise money for area nonprofits. Thousands of tourists, local residents, and—notably—farmworkers decked out in their finest clothes come to see the classic car and Chicano/a low-rider shows, to dance in the streets to music performed by live bands, and to treat their children to carnival rides, games, and maybe some strawberry pizza, kabobs, or empanadas.

Still, not everyone loves strawberries. American professional football player Tom Brady thinks they taste like "shit" (Burdge 2018). Indeed, the mass-produced berries we find at grocery stores don't have a whole lot of flavor. Contemporary plant breeding efforts have prioritized (and privatized!) color, shape consistency, and durability for worldwide shipment, high yields, and compatibility with pesticide and soil fumigant intensive methods of production (Guthman 2019) over and above taste. Yet people still buy them.

For a while, I couldn't eat them. On trips home to Pennsylvania, my mom would sometimes purchase California strawberries, from Watsonville, in an attempt to connect with me through my work. The things endured by the people I befriended, and my emerging knowledge of the human health burdens of toxic pesticide exposure made berries harder to stomach. My feelings about whether or not we should continue eating strawberries have shifted somewhat. I think other forms of activism and support for farmworkers are more important and urgent, albeit still imperfect, as I'll describe in the Conclusion to this book.

Strawberries are among many globally grown and distributed fresh agricultural commodities whose production is fraught with ethical and environmental issues. Growing strawberries requires significant, intensive, and extractive ecological management over land and natural resources and social control over farmworkers and markets. Farmworkers have a grounded and embodied knowledge of this, which is why some refer to strawberries as *la fruta del diablo*, the fruit of the devil. Picking strawberries requires workers to be *agochado* (stooped over) for hours at a time (Holmes 2013; Schlosser 2003) in often toxic and dangerous conditions. In stark contrast, strawberry marketer Driscoll's, describes the fresh berries it distributes and sells as "pure berry joy" and brands its allied teams of white-coated plant breeding scientists and mud-crust-booted agricultural field technicians as "joy makers" (Driscoll's, n.d.).

These complexities and contradictions involve significant conflicting human experiences and feelings, for the consumers of fresh produce, for the workers who tend, harvest, and pack fruits and vegetables, and everyone else involved. Having relatively easy and cheap access to food for some is made possible in large part by a multipronged "squeeze" (Copeland and Labuski 2013; Gálvez 2018; Guthman 2019) on and "cheapening" (Patel and Moore 2017) of farmworker communities and planetary ecologies in California, México, and beyond. We are implicated in this as consumers, but we don't have as much control over the food system via our purchasing power, and these truths about the foods we eat are rarely made transparent for us.

There are no easy answers to these dilemmas in our food system, and berries are no exception (Guthman 2019). This does not mean anthropological and activist engagement on these issues is useless, nor should we take things as they are as inevitable or normal. Talking through problems helps reiterate their realness and consequences. It opens up spaces for thinking about possible reactions and avoiding the comforting yet dishonest appeal of constant positivity even amid serious adversity and injustices (Ehrenreich 2010; Pavlovitz 2019), or the lure of quick fixes and universalist solutions that too often fail or create new problems of their own (Guthman 2019).

Fortunately, many doubts are being seeded about the unfulfilled promises of industrial food production. As more people struggle to deal with an array of intersecting social and environmental crises and inequalities, it has become harder to ignore the injustices built into our food system. These contradictions are, in part, what fuels the activism described in *The Devil's Fruit*. In Chapter 1, I sought to make farmworkers more familiar to us and to foster greater empathy for their daily lives, stories, and struggles. In this chapter and chapter 3, I aim to defamiliarize the strawberry as an everyday food object and as an example of the political arrangements and production logics supporting industrial monoculture: the growth of a single crop in mass quantities for the sake of market efficiency and maximizing profits, which often involves extensive resource extraction and heavy use of toxic synthetic pesticides and fertilizers.

"To make the familiar strange and the strange familiar" is a classic anthropological approach that aims to challenge and rupture what we think of and see as (ab)normal and (un)natural (Copeland and Labuski 2013; Myers 2011). In his satirical essay "Body Rituals of the Nacirema," Horace Miner (1956) writes about U.S. culture in the embellished and editorialized style used by anthropologists who traditionally studied and wrote about so-called exotic people in faraway places. Things like getting one's teeth cleaned at the dentist and following strict diet and exercise regimens to control the shape, size, and composition of our bodies (eating eight strawberries a day, for example) come off as less than rational when viewed through Miner's anthropological eyes and his playful yet serious subversion of the *etic* or outside researcher's perspective. Readers who aren't yet in on the joke see the Nacirema (American spelled backward) as an inhumane and grotesque tribe. When we realize Miner is writing about *us*, we are forced to reconsider what we think we know about our own culture and what we assume about the cultures of others.

In "following the object" and "making the familiar strange, and the strange familiar," I also build on archeologist Igor Kopytoff's "biography of things," an approach that "make[s] salient what may otherwise remain obscure" (1986, 67; see also Appadurai 1986). George Marcus and Michael Fischer's strategy of "cultural critique," the "disruption of commonsense, doing the unexpected, placing familiar subjects [in this case, strawberries] in unfamiliar or even shocking, contexts . . . to make the reader conscious of difference" (1986, 137), also facilitates my intentional reframing of strawberries and their ecosocially toxic production processes and labor relations. I see activist potentials in these ethnographic research and writing methods. They force us to question what is assumed to be inevitable, normal, natural, true, or even necessary in food systems and social relationships, including the routinized exploitation and harm endured by farmworkers and the use of toxic pesticides.

This chapter and chapter 3 defamiliarize the strawberry through a deep historical, legal, and ecosocial analysis of how strawberries are grown, where they come from, and what toxic products, processes, politics, and social relationships are involved in their production.[3] I show the multilayered consequences that the systemic squeezing and poisoning of land, resources, and im/migrant farmworkers in industrial commercial food systems has on farmworker health and welfare. Whether these harms are intentional or not should not excuse accountability and responsibility. I use ethnographic evidence collected from dozens of interviews and countless hours spent with farmworkers, participant observation and reaction via environmental justice activism, and a thorough assessment of gaps in occupational and environmental health safety policies in California and the United States to accomplish these goals. These two chapters are also myth busting exercises. Chapter 1 attempted to humanize and make more familiar the lives and stories of im/migrant farmworkers, who are routinely branded as alien, unwelcome, inconvenient, threatening, and dangerous. I critiqued the social construction of undocumented im/migrant farmworkers as mooches, malingerers, miscreants, subhumans, and not belonging in our communities. In these two chapters I address misconceptions about the purity and naturalness of strawberries and other fresh fruits and vegetables and see the monocultural landscapes produced by industrial agriculture as abnormal, dangerous, and almost extraterrestrial.

I present these "beyond the fork" (Saxton 2015a) views in the hopes that they instigate emotional and physical reactions among readers beyond merely changing consumption practices or never eating strawberries again (Guthman and Brown 2016a). Following the object as an activist and ethnographic research strategy can produce narratives, exposés, and images that disrupt deceivingly ordinary or benign "market values" and support ecosocial values of accountability, humanity, and justice (Krieger 2001). The evidence and creative responses I present support one of the overarching arguments of *The Devil's Fruit*: that industrial agriculture in its current forms is an example of a harm industry (Benson and Kirsch 2010). Acknowledging this truth, however harsh, inconvenient, or stomach churning it may be, is another first step in the challenging work of addressing environmental violence and social suffering in the fields and beyond and to imagining new agricultural and food futures.

THE INDUSTRIAL STRAWBERRY'S ROOTS

Strawberries, like all contemporary crops, from maize to mushrooms, have wild origins (Guthman 2019; Robinson 2013). Indigenous peoples in the Americas, up and down the Pacific and Northeast Atlantic Coasts, gathered wild varieties of strawberries, using them as food and medicine and in ceremonies. American Indians in New England continue to celebrate the Strawberry Moon, which marks the change in seasons and celebrates the arrival of the wild strawberries (Kavasch 1995; H. Murphy, n.d.) The highly anticipated seasonal arrival of wild berries of all kinds is still a significant cultural event in many other Indigenous and rural communities: from the Arctic tundra (Johnson 2014) to the radioactive woods around Chernobyl, Ukraine (Brown and Martynyuk 2016). The ecosocial relationships and processes involved in these harvests are markedly different from those of the industrial strawberry.

The union of the eastern (*Fragaria virginiana*) and western (*Fragaria chiloensis*) varieties of wild strawberries, which are small and soft, created hybrid varieties that have served as the foundations for contemporary horticultural (or farmed) strawberry breeding (Davidson 1999,757, Murphy, n.d.). During the colonization of the Americas and into the early twentieth century, settlers, naturalists, and missionaries studied plants; took rootstocks, seeds, and samples of all kinds; and experimented with them. Christopher Columbus and his contemporaries, for instance, were especially attentive to the market potential of the plants that were new to them in the Americas (Patel and Moore 2017, 50). Collecting and experimenting led to breeding, hybridizing, and mass-producing strawberry plants for home gardeners and nascent commercial growers. Breeders attempted to find the "perfect" strawberry suited for specific regions and with respect to ambitions to sell strawberries far and wide (Darrow 1966; Fletcher 1917, 157–182; Guthman 2019; Wilhelm and Sagen 1974). This proved no easy feat given geographically unique climates, pest problems, and varied access to distribution networks and markets across the United States and the world.

Early twentieth-century commercial California strawberry growers sought to capitalize on the state's deceptively bountiful natural resources to grow high-value fruits. They tried different varieties and growing methods with support from the then newly established land grant universities (Henke 2008). Specialty crops like strawberries, vegetables, and nuts would command higher prices on the mass market than grains, California's original "bonanza crop" that originally had attracted farmers and investors to the state (Wilhelm and Sagen 1974, 165–166; Stoll 1994). California's many microecologies feature different climates and soil types ideal for cultivating many different crops. On the Central Coast, year-round cooler temperatures, morning ocean fogs, and afternoon sunshine are good for growing things like strawberries, raspberries, blackberries, apples, leafy greens, and brassicas like broccoli and brussels sprouts (Guthman 2019; Wells 1996).

Historian Steven Stoll (1994) describes how early industrial growers in California framed these features as a "natural advantage." Early ideas about civilizing nature through farming reflect early California white farmer-settlers' values and socialization as entrepreneurs and venture capitalists (Patel and Moore 2017; Stoll 1998). But the expansion of industrial agriculture in California from the mid-1800s onward also involved significant human interventions, including the violent removal and slaughter of Indigenous and Mexican people and the appropriation of their lands for large-scale commercial farming and ranching. Settlers and investors sought ways to transform regional ecologies, including plants, aquifers, water systems, and landscapes—not to mention workers—to best support the production of massive quantities of food to feed a growing and westward-moving U.S. population. The generation of wealth as the lure of the California Gold Rush dwindled hinged on developing vast tracts of land for agricultural production. This fueled the development of infrastructure like ports, dams and irrigation systems, cities, railroads, and refrigerated shipping and transport networks to support California's expanding agricultural economy (Arax 2005; London et al. 2018). Most of these were also built by im/migrant laborers.

In these ways, strawberries and other industrially grown crops in California are part of the ongoing legacies of Manifest Destiny: a policy originating in the early 1800s that encouraged the westward migration of white settlers in order to facilitate U.S. political dominance through land grabbing, resource extraction, and intensive infrastructural development. Reaping respectable profits from fruits and vegetables required specialization or focusing exclusively on the production of a few crops on a grand scale, a practice known as monoculture or monocropping. Having large quantities of a single high-demand product made it more worthwhile to ship and transport agricultural goods like strawberries to urban markets where they commanded higher prices and higher returns in bulk shipments. This structuring of the food market encouraged farmers to grow monocultures over large areas, a practice that has intensified and expanded in California in the twenty-first century. It has also become an point of "fragility" for the industry (Guthman 2019) as monocultures possess and produce multi-layered vulnerabilities for people, plants, ecologies, and economies.

Monocultures shape landscapes, resources, and people. In her research on tea plantations in northern India, anthropologist Sarah Besky traces the colonial ideological roots of the monoculture and its human and ecological consequences:

> Champions of imperial expansion hailed monoculture as a triumph of science and technology over putatively wild landscapes and people. . . . Plantation work is equal parts caring and killing: pesticides and fertilizers, spades and sickles, irrigating water and combustible fuel. . . . Since plantation monoculture is an economy of scale, people and plants only need to be marginally healthy, marginally alive. *Quantity of life comes before quality of life. . . . Monoculture requires eradicating life, even though monocultures are often formed in the name of feeding or otherwise sustaining life.* The things that inhabitants of the industrialized North cannot live without—coffee, tea, sugar, bananas, as well as the rubber in our tires, the soy that binds dark chocolate, and the grasses that cover suburban lawns—are monocultured. (2017, emphasis added)

In their 1973 study *A History of the Strawberry,* University of California plant pathologists Stephen Wilhelm and James E. Sagan feature a photograph of a vast-reaching strawberry field outside Watsonville. The accompanying caption describes the transformation of the land, "from pasture and brush to a near-endless maze of strawberry beds" (Wilhelm and Sagan 1974, 229). Pasture and brush are coded as untamed and uncivilized, while strawberries, grown in expansive monocultures, are touted as more economically, ecologically, and socially efficient and aesthetically pleasing, like a well-groomed lawn. Those stands of pasture and brush themselves are legacies of other extractive and monocultural industries: early logging and cattle grazing.

De facto sentiments of Manifest Destiny remain embedded in agribusiness through the justification of the monocultural practices described herein, the highly concentrated ownership of land, and the feelings of entitlement to water and cheap laborers (Arax 2019; Farmview, n.d.; Macaulay and Butsic 2017). So, too, is the idea that industrial agriculture is an inherently benevolent and scientifically sound act of food production and land management and the only way to feed a growing population, reiterated in the mantras about feeding the world and creating jobs. The control over im/migrant growers and farmworkers is yet another form of ongoing extraction that intersects with the what Raj Patel and Jason Moore call the "cheapening" of nature, labor, care, food, and lives involved in industrial food production (2017). A hypervulnerable, malleable, and inexpensive group of diasporic and im/migrant workers—Chinese, Filipinos, African Americans, dispossessed whites, American Indians, Japanese, Punjabis, Mexicans, and Central Americans—maintain and harvest monocultures, historically and into the present. Synthetic chemicals, used to kill pests that threaten the viability of monocrops, which ironically provide insects with an easy-access smorgasbord, are the go-to method of ensuring crop profitability amid the unpredictability of nature and markets (Guthman 2019; Patel and Moore 2017).

The legacies of this history and its contemporary and highly toxic manifestations in the industrial strawberry fields harbor multilayered consequences for farmworkers', community, and planetary health: some are well established, others unclear or uncertain. Not knowing the relationship between a pesticide and a health outcome is routinely used as an excuse to foreclose preventative action when it comes to environmental health. To understand how farmworkers' lives and livelihoods are simultaneously made and unmade in the strawberry fields, we must first delve into the life and production cycle of the twenty-first-century industrial commercial California strawberry, which in its present iterations, from start to finish, hinge on toxic and synthetic chemical dependencies and human and ecological vulnerabilities.

STRAWBERRY LIFE CYCLES

Some crops, like lettuce and broccoli, are grown from seed. Modern-day strawberry varieties destined for commercial production are cloned in laboratories. The material from a strawberry "mother" plant produces numerous "daughters" that are replicated in specialized nurseries (Charles 2012). Desired genetic traits, such as pest resistance, high yields, texture, firmness, sweetness, color and appearance, transportation durability, shelf stability, and regional climate compatibility can be controlled for to varying degrees through plant breeding. This can be accomplished the old-fashioned way, through plant sex via cross-pollination, or via cloning, where tissues from plants with ideal traits are reproduced.[4] Ultimately, most industrial strawberry varieties are selected for their ability to produce consistent, thriving, and high-yielding crops that are also designed to thrive best in fumigated soil (Guthman 2019) and will arrive to grocery stores mostly uniform and unblemished. Sometimes this results in berries that are also relatively flavorless (Goodyear 2017).

Plant breeding research and development involves trade groups like the California Strawberry Commission, university researchers and extension agencies, berry grower-shipper companies and cooperatives, public-private research initiatives like Methyl Bromide Alternatives Outreach (MBAO), and state and federal government agencies. While varieties developed by the University of California remain in the public domain, accessible to anyone who wants to grow them, those developed by private companies—namely, Driscoll's—remain proprietary (Darrow 1966; Goodyear 2017; Guthman 2019).[5] Land grant universities in commercial strawberry growing regions (especially in California and Florida), along with pesticide companies and agribusinesses, have also invested a lot of resources and funding into plant breeding, the testing and development of fertilizers and pesticides, and nonchemical strategies and methods to boost yields.[6] Experimentation with soil fumigation dates back to the late 1800s, but widespread commercial use didn't begin until the 1960s (Guthman 2019; Newhall 1955; Taylor 1951; Warnert 2010a). Some landowners even capture higher rents on more recently fumigated lands, and crop insurance companies look favorably on soil fumigation as a way to ensure profitability, as crops grown on unfumigated soil are considered more of a financial risk (Guthman 2019).

Young strawberry plants, or runners, are fostered in nurseries in Northern California. A special class of highly experienced and mostly older farmworkers labor to pick off the flowers and green fruits to promote foliage growth (Bacon 2013). The soil used is chemically treated using highly toxic broad-spectrum soil fumigants.[7] This kills both beneficial soil organisms and harmful ones that spread plant diseases, like nematodes and verticillium wilt (a fungal infection that afflicts plant roots), which can wipe out entire fields if left untended (Guthman 2019). The daughter plants remain in the nurseries until they are mature, then transplanted to the coasts. When the weather warms, the plants are triggered into fruiting quickly and abundantly.

From August through October, the end of the Pájaro and Salinas Valley harvests, strawberry fields are prepped for the following year. The soil is chemically treated with fumigants every few years, especially if previous crops showed any signs of disease. To begin, a crew of farmworkers rips up the old plastic ground covering used to prevent weeds, manage moisture, and protect berries from touching the ground where they are susceptible to rot and bacteria. In the process, dirt, fertilizer, and residual pesticide dust are churned up, breathed in, and carried home on workers' clothing and shoes. One farmworker described how the chemicals "rust" the plastic, eating holes in it and changing the color from clear to a dusty orange. Tractors till the old foliage under and level the ground. Then, typically at night when wind and temperatures are stable, spaceship-like machines insert long shanks into the soil, injecting gas or volatilized liquid fumigant pesticides, killing soil-borne pathogens and other organisms. Simultaneously, long plastic tarps, either virtually or totally impermeable film, are rolled out to seal in the fumes. In

FIGURE 3. Ripping up last season's plastic tarp. (Photo courtesy of Amadeo Sumano.)

FIGURE 4. Heading home. (Photo courtesy of Amadeo Sumano.)

essence, this is a very high-tech Saran wrap that is supposed to keep fumigants sealed in to do their work and prevent drift to nontarget areas and communities.

The plastic tarps are not foolproof. They can tear, and undulate in the wind, exposing people at other worksites and area homes and schools (Guthman 2019; J. L. Harrison 2006, 2008, 2011). Fumigation season in the Pájaro and Salinas Valleys coincides with children returning to school. Even when precautions are taken, applicators, farmworkers, and other passersby can still get sick from exposure during or shortly after applications. Longer-term consequences are not well studied because the models and tools for such research remain underdeveloped and underfunded.

The California Department of Pesticide Regulation maintains the California Pesticide Illness Query, a searchable database of *reported* exposure incidents. Between the years 1992 and 2016 in Monterey and Santa Cruz Counties, there were 215 incidences of pesticide drift from soil fumigant applications alone affecting farmworkers. During the same time period, 383 drift incidents affected nearby residents at homes, schools, and businesses (CA DPR, n.d.). Affected farmworkers and communities experience health effects ranging from headaches and nosebleeds to chronic and permanent respiratory and neurological problems (Environmental Working Group 1997; Quintero-Somaini and Quirindongo 2004).

Before, during, and for a period after a soil fumigant or other pesticide application, telltale red-lettered signs adorned with skulls and crossbones must be posted at the peripheries of recently treated fields. Letters may also be posted at nearby residences. Increasingly, public schools are requesting advance notice of pesticide applications. Signs warn passersby to stay away, in English and Spanish: "Danger/Peligro." Most residents and workers have nowhere else to go. Other nontarget organisms may also be affected by soil fumigants, including species that are beneficial in agriculture (Guthman 2019; Ibekwe 2004; Saxton 2015c).

After forty-eight hours, and sometimes up to two weeks (depending on the chemical, the application method, the time of day, and the weather), it is assumed that the levels of soil fumigant gasses and vapors in and around the fields pose no harm. Farmworkers score and rip up the plastic film by hand; they wear no protective gear. Professional applicators are equipped with spacesuit-like 3M brand white zip-up hooded coveralls and sophisticated full-face respirators. Efforts to develop stronger tarps to mitigate drift have a hidden consequence: keeping the gasses under wraps means that when workers rip up the tarps to prepare the fields for planting, they can be exposed to higher concentrations of fumigants (Gao et al. 2011; Guthman 2019). Gerardo, a farmworker whose story is recounted further in chapter 5, remembered what it was like to remove the tarps. The gases released always burned his nose and eyes.

FIGURE 5. A recent soil fumigant application and its telltale warning sign. (Photo by the author.)

Used tarps are then taken to a storage site where they'll remain until they can be recycled. I found out about this relatively hidden part of the production cycle from resident environmental activists. They started noticing piles of baled plastic tarps, mulch, and drip irrigation tubing, mostly from strawberry farming, accumulating behind a gated and fenced property. For years the spent plastic sat, in rain, wind, and sun, releasing lingering dusts and residues into the soil, air, and water.

Once the fumigation process is complete, the fields are prepared for planting in November and December. Tractors form uniform mounded rows, approximately eighteen inches high. A tractor rolls out thin strips of perforated black drip irrigation tubing, or "tape," which helps conserve water and protects plants from mildew, mold, and fungal growth. Drip irrigation systems are also used to deliver synthetic fertilizers and pesticide treatments, including some soil fumigants, directly to plants' root systems. Each row is topped with plastic mulch, wound out from large rolls off the back end of another tractor. Farmworkers weigh down the mulch with shovels full of dirt.

After the fields are prepared, chemically and otherwise, farmworkers hired at an hourly rate carry boxes of new strawberry transplants, raised in fumigated and sterile soil (Guthman 2019), recently arrived from the colder northern nurseries.

FIGURE 6. The dump. (Photo courtesy of Timothy Flynn.)

Walking down the rows they make a series of small punctures in the plastic mulch, each of which is filled with a single young strawberry plant. More and more, this is being done with machine labor guided by workers. Careful and even spacing allows for bushy growth when the temperatures increase. Young strawberry plants may be treated with fungicides and insecticides as well as extra doses of fertilizers to help ensure survival through springtime. By April and May, warmer weather instigates flowering and fruiting. In the Pájaro and Salinas Valleys, the harvest season lasts from late April through early October. Along the more southern coast near the communities of Oxnard and Santa Maria, and farther south into México, harvesting happens in the winter months. Some growers are using plastic greenhouses to extend their growing seasons. Others go global. Grower-shipper company Driscoll's engineered the year-round berry growing cycle by expanding its production and marketing to places like México and Morocco. They have played a major role in cultivating our desires for fresh fruit on demand (Goodyear 2017). Companies, not consumers, have engineered and driven the demand for strawberries.

While strawberry plants are actively producing fruit, more pesticides, including herbicides, fungicides, and insecticides, are applied in or near the fields, either from the ground, with tractors; manually, with hand pumps; or aerially, from helicopters. In the fields, farmworkers are exposed to residual pesticides on plants, or from those that drift from other fields through the air. Dusts and residues follow them home on their clothes, where children are secondarily exposed (Bradman et al. 2007; Quirós-Alcalá et al. 2011).

FIGURE 7. Transplanting strawberries. (Photo courtesy of Amadeo Sumano.)

Come harvest time, farmworkers quickly make their ways up and down the rows, pushing aside foliage in search of berries that are not too pale and not too ripe. A deft flick of the wrist, which takes a lot of practice and skill to master, releases the fruit from the plant without any green stems intact and without bruising or smashing the fruits. Workers labor for hours at a time in an excruciating hunched posture, which they refer to as *doblado* (bent over) or *agachado* (stooped, both body and soul) (Holmes 2007). To facilitate simultaneous and

FIGURE 8. A worker covered in strawberry juice and pesticide residue.
(Photo courtesy of Amadeo Sumano.)

FIGURE 9. Farmworkers and an active pesticide application's drift.
(Photo courtesy of Amadeo Sumano.)

efficient picking and packing into the clear plastic clamshell containers, workers use low-to-the-ground carts known as *carretillos* (wheelbarrows). Each cardboard flat fits about twelve clamshells holding a pound of berries each, or six two-pound clamshells for orders destined for warehouse stores like Costco.

Berries must be picked and arranged quickly and artfully into the baskets. Speed is a necessity when one is paid by the piece rate, or per quantity of flats filled. At the time of my research (2010-2012), the piece rate in the strawberries according to farmworkers ranged from $1.00 to $1.25 per flat, and for higher value raspberries or blackberries, $2.00 to $2.50 per flat. Organic products may fetch a slightly higher piece rate, but many workers complain about the lower yields in some industrial organic commercial fields.

Upon filling a flat, farmworkers walk, or sometimes sprint, up and down the rows in attempts to maximize their time picking. A *ponchadora,* or checker, will inspect farmworkers' harvests, as may a foreperson and supervisor. If a worker can't keep up with other members of the crew, he or she is asked not to come back. There isn't much time or patience afforded for newcomers to learn or adapt.

Once the day's quota has been harvested, packed, and inspected, another crew of workers arrives with a truck equipped with a long bed that the flats are strapped to. From there they are transported to the coolers. After a thorough chill, the berries are distributed to wholesalers and retailers all over the world.

But before strawberries make it to market, they are dosed with fumigant pesticides once more. Large coolers are pumped full of gas meant to kill off any lingering bugs, fungi, mold, mildew, and/or bacteria. The U.S. Department of Agriculture and similar agencies in other countries mandate postharvest and preshipment fumigation and quarantine protocols to prevent invasive pests from

FIGURE 10. Stooped labor. (Photo courtesy of Amadeo Sumano.)

spreading, thus devastating industrial agricultural production nationally and globally. Following a preshipment fumigation, the invisible gases are released into
the atmosphere. These, too, can drift to residential communities, schools, and
other farm worksites. An activist friend who lived upwind of a major cooler
described how the position of her house on the rim of a small canyon channeled
the pesticide drift straight in her direction on strong winds blowing inland from
Monterey Bay.

FIGURE 11. *Carretillos* (wheelbarrows) lined up at the ready.
(Photo courtesy of Amadeo Sumano.)

FIGURE 12. Packing and inspecting berries. (Photo by the author.)

At supermarkets, strawberries and other berries are arranged in abundant displays, adorned with the carefully designed logos of marketing companies like Cal-Giant, Driscoll's, Good Berry, Limited Edition, and Well-Pict. Each of these companies is furnished with fruit through affiliated grower-shippers (like Reiter Affiliated Companies for Driscoll's) or a series of growers or smaller-scale sharecroppers who have a contract with a larger company (Guthman 2019; Wells 1996). Consumers may or may not pay much attention to the labels or brands, but the lure of fresh berries in January is often too tempting to resist.

STRAWBERRY FIELDS AS EXTREME ENVIRONMENTS

When conjuring images of extreme and polluted environments, people may think of barren, desolate, almost lunar landscapes. Some so-called natural environments, such as the Arctic, the top of Mount Everest, or the Sahara, are culturally classified as extreme because of the special challenges they pose to sustaining human and other forms of life. These might include industrial zones, such as the oil refineries and chemical plants in Louisiana's Chemical Corridor (Singer 2011), along the New Jersey Turnpike, or outside Houston, Texas (Harper 2004) and Richmond, California (Ferrar 2017). Their mazes of tubes, tanks, and billowing smokestacks release hazardous emissions and ominous flames within walking distance of residential communities. Others might envision the burgeoning fracking lands, where oil and natural gas extraction leaves indelible marks on the land and contaminates air and water with synthetic chemicals used to fracture solid layers of rock to access reserves ever deeper in the earth (Wylie 2018). Perhaps they envision former nuclear testing zones and other accident sites like Los Alamos, New Mexico, or Fukushima, Japan.

While performing a search on Google Earth for satellite pictures of oil fields, artist Mishka Henner (2015) came across high-resolution photographs of cattle feedlots in Texas. He artfully stitched them together in an effort to better visualize their environmental impacts on the landscape. The visual residues and landmarks from the wastes of confined animal feeding operations, fossil fuel extraction, and other similarly intensive activities look eerily similar. All are products of human-driven environmental transformation and destruction.

Medical anthropologist Valerie Olson (2010, 2018) defines "extreme environments" as spaces where human, technological, and chemical interventions are used to observe and control life. In her research on the National Aeronautics and Space Administration's space biomedicine program, she observed how the agency's doctors, responsible for monitoring and attending to the health of astronauts, naturalize the extremes of the outer space environment and its varied and at times harmful impacts on the body. As Olson explains, "In extreme environments . . . investments of power and knowledge shift from life itself to the sites of interface among living things, technologies, and environments" (2010, 170). This entails a suturing together of health and environmental governance and norms (or assumptions about what's best) wherein biomedical and technical models help normalize states of being, health, and illnesses in certain bodies, environments, and contexts. Doctors monitoring astronauts use the concept of "space normal" (Olson 2010, 170–171) to differentiate between health problems of concern and symptoms that are deemed reasonable risks or facts of life for those living and working in outer space. Risk thresholds of stress or exposure are thereby defined in accordance to models that aim to predict how the generic human body will react and change during different phases of a space mission (Olson 2010, 183).

Similar logics operate in the production and maintenance of monocultural strawberry fields and the management of workers. This includes the use of probabilistic risk and computer models that attempt to predict how effective pesticides are at killing target pests and how they might affect the health of people and other organisms in surrounding environments (Froines et al. 2013; Peña 2011). Neither environmental health risk assessments and governance nor decisions about how to run agribusinesses or manage workers are politically neutral. Herein is another dimension of the monoculture noted by Besky: "single-species landscapes are among the most vulnerable to blights, diseases, and droughts," for both the plants and the people who work them (2017). Thus, monocultures, with their promises of productivity, efficiency, and profitability, fail at sustaining life, inclusive of the farmworkers who tend to them.

Even seemingly benign introductions of technologies framed as improving worker productivity (and potential pay) may engender hidden harms. For example, some strawberry companies are using diesel-powered tractors that workers trail as they load their harvests; an electronic scanner keeps track of the number of boxes picked.[8] Workers following mechanized harvesting rigs are expected to pick even faster since they no longer have to walk or run back and forth to check in their harvests. This also increases the time spent stooped over, however, heightening the risk of repetitive stress injuries and strains, falling off rigs, being run over, and inhaling fumes and dust (Fathallah 2010; Horton 2016b). The threat of further mechanization, to the point where farmworkers are eliminated almost entirely, also looms as a means of social control (Seabrook 2019), although growers see it as a way to adapt to labor shortages, and rising wages and costs of doing business tied to environmental health and safety regulations (Guthman 2019).

Some laud these and other forms of agribusiness paternalism as a good-faith effort at demonstrating care and concern for worker welfare and as a sign that agribusinesses are starting to make substantial investments in farmworker wellness. Many of these discourses rationalize good treatment, good wages and benefits, and good jobs as good business sense. Capturing worker loyalty and keeping workers healthy and injury free saves businesses money. Well-intended paternalism—however insufficient it may be—can also perpetuate worker silences through fostering feelings of gratitude for what one has: a job, a nice boss who sometimes buys lunch, and so on (M. López 2011). Silences also come from more violent forms of social control. Sexual harassment and assault, rape, wage theft, gendered and racialized antagonisms, and threats of firing or deportation are also routinely used to coerce and control workers in the fields (Castañeda and Zavella 2003; Yeung and Rubenstein 2013). Silence also comes from desperation—most workers don't have many choices about where they will work or under what conditions. They are often at the mercy of farm labor contractors (FLCs), third-party employers that growers and grower-shippers use to avoid responsibility for worker health and safety.

The voices of caution and embodied experiences or scientific observations about the human and ecological harms of industrial agriculture do not always get heard at the same volume as the concerns of agribusinesses and markets (Froines et al. 2013). To accept these tactics—be they forceful or gentle (Burawoy 1976, 1979)—as normal or natural, or even logical given the unideal circumstances and market pressures growers are facing (Guthman 2019), overlooks "monoculture's power ... to breed not just a dearth of biological difference across landscapes, but a creeping in-difference to the radically uneven impact of capitalism on ecologies, identities, and planetary life" (Besky 2017). It is to the lives interrupted by monocultural cultivation and labor practices in industrial agriculture that I now turn.

LIVES INTERRUPTED

Like that of strawberries, the cultivation of raspberries and blackberries (collectively known as caneberries for their woody, thorny stalks) in the Pájaro and Salinas Valleys has transformed landscapes, ecologies, economic opportunities, and people's bodies. Workers often expressed that working with *las moras* (their general word for caneberries) was an improvement compared to *las fresas* (strawberries). Caneberries grow "up," climbing specially made trellises. Farmworkers harvest them from more upright postures and can change positions more regularly, moving up and down the plants vertically to seek out the best berries, whereas in the strawberry fields, workers are almost constantly bent over.[9] Raspberries and blackberries also earned farmworkers higher piece rates, of two to three dollars per flat of twelve clamshell containers in 2010. The fruits are more delicate, requiring extra care during harvest and hand packing to avoid bruising and smashing. Piece rates were even higher for organic berries, which command even better prices on the market; an eight-ounce clamshell of these can sell for six to seven dollars.

Most commercial caneberries are grown under tunnels: arches covered by large swaths of plastic sheeting. A librarian in Watsonville, curious about my research, asked if the tunnels helped protect workers from the sun. In reality, they protect and promote the growth of the berries. It gets *really hot* inside the tunnels, which helps extended the fruits' growing season, keeping them warm and dry so that they can grow nice and plump. Tunnels also keep the dust down and prevent birds and other pests from overzealous feeding.

I had a chance to observe the work of blackberry harvesting in 2009, when a grower I approached granted me permission to hang around so long as I didn't interfere with the work. Dust, a combination of dried dirt and pesticide and fertilizer residues, swirled in the hot air, churned up by the farmworkers' brisk inspections and shuffling from plant to plant. Men and women labored side by side, standing and stooping, up and down, row after row, gently parting the foliage of each plant, searching for ripe (but not too ripe) fruit, and deftly sorting and packing berries on the spot. Those that felt too soft between the fingers or that showed signs of insect or worm damage were tossed to the juice-stained ground. Prime berries got deposited into plastic buckets strapped to farmworkers' bodies with leather holsters worn around the shoulders or at the waist. Everyone had his or her own packing and sorting tables, crafted of plywood and affixed with workers' numbers so that their harvest quotas wouldn't be mixed up. Upon filling a few buckets, everyone carefully and neatly arranged the berries into prelabeled containers, which in this instance also happened to feature a "certified organic" seal. Some soft berries might also get sorted into a *jugo-* or juice-grade bucket, which would be sent to companies that made drinks, jams, and fillings.

I certainly was not exerting myself on this unusually hot day. I slowly paced down the middle of the rows or stood at the field's edge with my notebook and camera in hand: close enough to document the work as the harvest unfolded without being in the way. Nevertheless, I felt the heat getting to me. The inner brim of my hat became soaked with sweat. I felt dehydrated. I was barely moving.

Abruptly the work stopped for lunch. Farmworkers offered me one of the plastic lawn chairs that employers are required to provide at California farm worksites for use during breaks. These, along with access to potable water, gender-segregated bathrooms, shade, hand washing stations, and legally mandated time to rest and eat represent yet another series of long-fought worker safety and welfare victories. Unfortunately, they aren't always enforced or guaranteed.

I sat to chat with the crew. Folks were curious about who I was and why I was there. A woman crew member limped over to the rest area, looking visibly pale and sweaty. She grasped her right side in discomfort, wincing a bit as she walked. I asked what was wrong, and the group exclaimed in unison, "¡Está embarazada!" (She's pregnant!) She sat down, guided by the support of a *compañera* into a chair, and drank a little bit of water.

In the spirit of hospitality, workers offered me a nectarine and an apple. I remember feeling guilty for accepting such gifts, not fully understanding that from then on, no matter where I went or whom I visited, people would give me fresh fruit, drinks, tamales, and even entire meals as a sign of welcome and respect. I came to embody this trust, gaining twenty-five pounds during the course of my field research.

Naively, I suggested that maybe their *compañera* should eat the fruits instead. The workers explained to me that, in general, no one ate or drank very much at work. The heat and constant intensity of physical work, standing up and down, bending and unbending, sometimes induced nausea and vomiting. An empty stomach stifled this very uncomfortable side effect, but it likely caused other problems. Even though California law mandates lunch breaks whenever someone works at least five hours, the nature of farmwork itself limits whether or not one can really take advantage of these rights. The need to work long hours at a quick pace also discourages workers from taking breaks, even when it is extremely hot, they need a drink of water, or they feel sick (Horton 2016b). A loophole in the policy is that no rules *require* employers to make workers to stop working during breaks; they only have to offer the opportunity (FELS 2014, 49–50).

The manual dexterity and repetitive motions in all kinds of berry harvesting often leads to debilitating and painful musculoskeletal injuries, as the experiences of farmworkers in chapter 1 showed. Indigenous farmworker Yolanda worked in the raspberry fields right up until the birth of her son, Reynaldo, in late October. The harvesting season had just come to a close. At that time she lived in what amounted to a closet in a house with five or six other families. Social workers told her that if she wanted to keep her baby, she would have to prove that she could provide a safe home for him. Her coworker and *compadre* (godfather to her son) commented to me that he never saw anyone work so hard and fast as Yolanda did.

In our conversations, which usually took place during car rides to church events, the *tianguis gratis* (flea markets) at the Center for Farmworker Families, and various appointments, Yolanda recounted that she came to the United States to escape an abusive boyfriend, to be able to provide for her son, and to help out her recently widowed mother and younger siblings. Yolanda's father had migrated back and forth between California and Oaxaca, and in California between the vineyards of Santa Rosa and the strawberry fields of Oxnard and Watsonville. He died from complications of diabetes around his fifty-fifth birthday. These multilayered circumstances did not allow Yolanda to take any leaves of absence from work, not even when she was pregnant. After Rey's birth, Yolanda found a wintertime job in a flower nursery. Her younger sister, Yesenia, then fifteen, stayed home to care for Rey. By Rey's fourth birthday, which marked Yolanda's twenty-seventh, her fingers, hands, wrists, and shoulders felt chronically numb, stiff, and achy: possible signs of carpal tunnel syndrome.

Technically, California law allows for pregnant workers, irrespective of citizenship status, to apply for up to four weeks of paid disability leave before delivery, in addition to unpaid leave, without risking job loss. Maternity leave for farmworker women is of special concern given the occupational hazards posed by pesticide exposure, which include miscarriages and stillbirths, a number of birth defects, and developmental delays (Migrant Clinicians Network, n.d.). Another farmworker *compañera*, Lizbeth, gave birth to Yuleni, only to find that her daughter had a severe congenital heart problem that would require around-the-clock care. Lizbeth worked in the fields through her pregnancy but quit her job so that she could accompany Yuleni to appointments at a university hospital and to be available whenever the visiting nurses came around for home checkups. I met one of Yuleni's nurses, Rachel, during my own home visit. Rachel had countless patients, the children of farmworkers just like Yuleni, and suspected that pesticide exposure contributed to the numerous birth defects she observed during her rounds in the Pájaro and Salinas Valleys.

Some agribusiness companies offered pregnant women additional maternity leave benefits that exceed those required by law. Other farmworkers referred to informal systems of cooperation among themselves as a means to ease some of the burdens of farmwork for pregnant *compañeras* who had to keep working—namely, delegating the heavy lifting requirements to male coworkers or helping sick, pregnant, or elderly workers finish their rows and meet the quotas. In some cases, other family members picked up the slack while their wives and mothers took time off. Indigenous farmworkers Martín and Veronica were expecting their sixth child. When I interviewed them, they knew plenty about pesticides from their experiences working in the berries and from helping their families back in México maintain traditional pesticide-free *milpas*, fields where corn, beans, squash, and wild greens grow together in a supportive interspecies system. These polycultures are the exact opposite of a monoculture. Martín and Veronica came of age at time when many west central Mexican agricultural communities faced mounting pressures to switch to chemically intensive hybrid and genetically modified varieties.

During my visit to their home outside Salinas, Martín and Veronica's two oldest daughters asked me to donate some funds to help support their *tía* (aunt), who also worked in the strawberry fields. I deposited some cash into a cardboard box, adorned with a blurry cell-phone photograph of a stillborn baby: eyes closed, body lovingly swaddled. This was their *primo* (cousin), and their aunt wanted to send his remains to México to be buried alongside other relatives. Stories of miscarriages and stillbirths abounded in my conversations with farmworker women, who sometimes, like Lilia (in chapter 1), counted miscarriages or stillborn babies as parts of their family tree. Some contemplated how pesticides and stress at work and at home affected their reproductive health, while others resigned it to God's will (*si dios quiere*).

For her sixth pregnancy, Veronica took an extended disability leave. Martín looked visibly tired after coming home from work. Nevertheless, after a dinner of lentil soup and tortillas, he carried on into the evening hours, sorting a large box of freshly picked organic beans that someone had given him. He would sell these to his extended family and neighbors to help make ends meet.

Growers and contractors, and sometimes farmworkers themselves, rely on logics that racialize and gender certain kinds of farm and nonfarm labor, including the delicate work of raspberry and strawberry harvesting (Benson 2008a, 2008b; Cartwright 2011; Holmes 2007, 2011, 2013; Horton 2016b; Figueroa Sanchez 2013; Zavella 2011). A grower for a larger berry company, whom I met on a farm tour, emphasized his farmworker hiring preferences to a group of junior college students studying organic horticulture: "The shorter the people are, the higher the beds, the faster you can pick." Their professor described another grower who used a doorway five feet in height to help screen prospective workers.

These conflations of race with innate biological traits and abilities, while baseless and extremely racist (Holmes 2007, 2013), justify the continued institutionalization of such employment practices. They also heighten workers' vulnerabilities to work-related injuries, pesticide exposure, and sexual and economic exploitation. While farmwork is ascribed as natural or innate to Mexicans, the ability to live and work to industrial agriculture's standards of hyperefficiency and productivity are impermanent and temporary for many workers. Younger workers are not immune. Women Yolanda's age and older start wearing wrist, knee, and back braces as a means of self-care to try to avoid clinic visits and to keep their jobs. Still, some end up permanently disabled before their fortieth, or even thirtieth, birthdays, as was the case for Milagro (in chapter 1), and Armanda and Aniceta (see chapter 4).

In reality, many workers don't feel like they have a choice to privilege their health over their wage-earning potential. Even the opportunity to eat and drink, to replenish their energy, is mitigated by how working hard in extreme conditions makes farmworkers feel. Some workers told me they just ate a piece of fruit or drank Ensure (a meal replacement shake), soda, Gatorade, or energy drinks to get through the day without feeling sick. They'd relish heartier meals at home with their families: rice and beans, fatty cuts of meat, *caldos* (broth-based soups) of chicken or beef and vegetables, scrambled egg–based dishes, all served with plenty of chiles and tortillas: something to look forward to.

ARE MARKET VALUES HUMANE VALUES?

This layered accounting of farmworker families' fraught and devastating experiences with biological and social reproduction illustrates more of the hidden disruptive and deadly consequences of monocultures from an ecosocial perspective. These experiences don't get captured in the California Occupational Safety and Health Administration injury rate data (Horton 2016a, 2016b) or pesticide reporting records that are the official mechanisms for keeping track of occupational injuries and harm. In normalizing the ecologically and socially toxic environments of strawberry fields and minimizing or ignoring farmworker and environmental health consequences tied to industrial agriculture, pesticide users, regulators, researchers and developers, and manufactures assume predictability, inculpability, and ultimate expertise over ecosystems and bodies (Olson 2010, 185); that there is such a thing as "acceptable risk" when it comes to the trade-offs produced by industrial agriculture and absorbed and embodied by farmworkers (Peña 2011, 212). The toxic, extractive, and exploitative ideas and systems governing strawberry production are normalized as what's necessary to feed the nation and the world and provide jobs to people who are deemed otherwise unemployable.

Monocultural and market values that sustain industrial strawberry production and create jobs for farmworkers are often privileged, in policy and grower-shipper circles, over human and ecological values concerned with the effects of pesticides and labor exploitation on farmworker lives and health. Julie Guthman (2017) has argued that environmental justice groups involved in antipesticide organizing will fail to make significant headway in their efforts to reduce harm to farmworkers if they don't attend to industry's evolving concerns about labor shortages. Organic and other less chemically intensive farming methods provide even more jobs that will attract farmworkers because they are much more labor intensive, requiring hand weeding and tending of the plants and land (Guthman 2004, 2017).

Meanwhile, labor and environmental policies are being rolled back through vocal support and silent complicity from the agricultural industry, which purports to care about worker health and welfare, at least as far as it relates to industry's need for workers to harvest crops and turn a profit. Indeed, from both market and ecosocial standpoints, "it makes little sense for industry to treat farmworkers as disposable by neglecting the health effects of pesticides" (Guthman 2017, 16). But the claim that "it makes little sense for activists to dismiss industry concern with farmworker employment when a shortage provides a new platform for making the case against fumigants," and for more labor-intensive nontoxic farm jobs, does not seem grounded around industry and state responses (or lack thereof) to date (Guthman 2017, 16). Nor does it acknowledge activists' often close relationships with farmworkers (sometimes they are related to them), their intimate understandings of the relationships between lives and livelihoods, and their quests for new ways of living and working in the world that won't kill and maim their friends, relatives, and neighbors, and maybe will even leave them with fruitful legacies and businesses to bestow to future generations.

Industry responses to labor shortages in California in the wake of the administration of President Donald Trump's crackdown on undocumented immigrants have to date involved substituting temporary H-2A workers for U.S.-based Latinx im/migrant farmworkers. Resident farmworkers are either aging, both naturally and prematurely (through chronic occupational injuries, diseases, and stresses), have left during other precarious periods of job shortages (such as with droughts), and/or are being deported. Efforts are underway to reduce rights and protections for both resident farmworkers and temporary workers, too.

In addition, many California agribusinesses are moving toward mechanization in the fields and shifting to crops that require less labor because wages are rising and regulations are becoming more costly to implement (Guthman 2019, 144). Eliminating the need for resident, im/migrant, and contingent labor reduces the associated costs—such as wages, insurance, equipment, training, and meals and housing in the case of H2-A temporary farmworkers—of recruiting and maintaining a workforce. This is by no means a new movement or process. Mechanized harvesting equipment for fresh fruits and vegetables is in use for a number of crops, including tomatoes, lettuce (Friedland 1981), almonds, and wine grapes. It can significantly reduce the numbers of paid workers required for harvesting, and some argue that this opens up new opportunities for training farmworkers for more technical and higher-paying work operating or maintaining the machines. There are also less-than-humane motives at play, however. In 2013, I attended the annual MBAO conference in San Diego as an observer. I listened as one prominent researcher posited that harvesting machines, while initially expensive, aren't affected by pesticides. They don't require expensive workers' compensation coverage and ultimately don't need breaks, overtime pay, holidays, or even wages. It is a major area of research and development for the California Strawberry Commission (n.d.-a) and companies like Driscoll's (Guthman 2019, 144).

These are the very things labor activists and workers have been fighting for centuries and continue to demand across occupational sectors. Guthman's argument does not attend to the tensions and complexities that farmworkers endure in their homes, their work lives, and their fraught relationships with growers, supervisors, and farm labor contractors, or to the long history of violence and blacklisting against farmworkers and activist allies who make demands for better living and working conditions. I also think it falls far short of acknowledging how many environmental justice organizations take part in the imaginative and on-the-ground labors of building alternative agricultural futures that are more inclusive, intersectional, and just for farmworker communities (Burke and Shear 2014; Myers and Sbicca 2015).

While robotic harvesters may still be in the research and development stages for delicate crops like strawberries, some growers, agribusiness industries, and researchers are framing mechanization as a good and desirable solution to the labor problem. This represents another dimension of monoculture, where only certain lives and livelihoods are valued and sustained, even, at times, to the detriment of some growers who are going out of business amidst labor, land, and competitive pressures (Guthman 2019). Others suggest that if the trend toward farm mechanization is more fully realized, farmworkers may benefit little from working on commercial-scale industrial farms, whether they are organic or not. As Seth M. Holmes (2013) observed during his fieldwork in Washington State with Indigenous farmworkers, organic blueberry fields, which would have given workers jobs with less exposure to pesticides, were being mechanically harvested. Meanwhile, workers were relegated to picking exclusively in the conventionally grown fields. Similarly, workers formerly employed in other industries have not fared well following shifts toward mechanized production—for example, in coal mining, the automotive industry, and the textile industry. Not much has been done to retrain and reassign them, despite promises to do so.

The lingering silences accompanying large companies' ongoing neglect of farmworkers' workplace health, welfare, and wage concerns in both México and the United States are not so easily reconciled through fair and balanced conversations about how activists and growers can work together, especially when the power dynamics are so uneven. While growers may feel powerless and hopeless given market pressures, unwieldy political shifts, and changes in the weather, they do have some protections at their disposal that farmworkers do not. Crop insurance and access to federal aid in the event of major disasters like the wildfires, droughts, and hurricanes of the last few years are not afforded to farmworkers who are not U.S. citizens. Injured citizen farmers are eligible for Social Security Disability Insurance, but undocumented farmworkers are not. Indeed, many farmworkers have been pressured to keep working amid hazardous and smoky conditions despite laws that are supposed to prohibit this (Herrera 2018). Companies like Driscoll's, among others, have billions of dollars at their disposal that they could share profits more equitably with the growers and farmworkers who make their business succeed. Some large companies, like D'Arrigo Brothers, a major lettuce harvesting firm, have taken steps to recruit and retain skilled workers, creating attractive benefits packages and offering higher wages in collaboration with the United Farm Workers (Sherman 2018). Even if individual growers cannot, at the very least agribusinesses and grower-shippers could realign their research, development, production, and labor models to more proactively investigate and prioritize nontoxic and healthy working conditions and they could take more active political stances on protections and support for immigrants. State and federal governments could also do more to prioritize and incentivize just and ecological food systems. But, in general, they do not.

I have presented an ecosocial critique of the strawberry industry that some may find harsh and unforgiving. While "in capitalist economies, making a living is a necessity for practices that ensure at least a modicum of health and that modicum of health is necessary for making a living," current "concerns over farmworker livelihoods" (Guthman 2017, 2), as expressed and framed by industry and in the work-driven culture that drives much of human existence and survival (Besky and Blanchette 2018) merit further concern and skepticism. This is especially true when we consider the environmental health effects of monoculture from pesticides and other occupational injuries and illnesses and their potential to continue inflicting harms intergenerationally and transnationally (Besky 2018; Saxton and Stuesse 2018; Unterberger 2018). At stake are limits on the lives of farmworkers and their children, grandchildren, and great-grandchildren.

Kregg Hetherington writes about the conflicting life and livelihood values of monoculture in the context of industrialized soy in Paraguay, a country where peasants have been systematically displaced from land and livelihoods. Through producing commodities and garden crops with the support and encouragement of the state, these small farmers became "citizens capable of expressing the country's sovereignty over its territory" (2016). Mechanized and industrially produced soy is robbing them not just of land, but of identities and opportunities.

"Agricultural labor for *campesinos* [peasant farmers and farmworkers] is not just a living, but a qualified form of living.... Soy farmers are not ... uncaring about human life beyond their own.... The difference is that theirs is a care for human life in the abstract, a globalized bare life that specifically denies the national difference that *campesinos* hold dear" (2016). There are parallels in this history of industrial agricultural transition that mirror those endured by Mexican farmworkers, many of whom were once small-scale farmers themselves before NAFTA destroyed Mexican rural communities (Galvéz 2018). Thus, the moral economies of caring—or *making* people care—within industrial strawberry growing aren't so easily reconciled by capitalist values of jobs and economic growth, labor crisis or not. Cultural changes in agriculture will require more than industry partnerships and compromises, especially when we consider the power and resource disparities between industry and communities in California and transnationally throughout the Americas.

We cannot forget that while the hardships and challenges of California growers and farmers are real and interconnected to the health and welfare of farmworkers (Benson 2008a, 2008b; Guthman 2017, 2019; Holmes 2013), they are at the same time "unequal to im/migrant workers' vulnerabilities, especially when it comes to . . . the long-term consequences of bodily harm" caused by occupational injuries and illnesses, exposure to toxic substances, and im/migration-related injustices (Saxton and Stuesse 2018, 69). Similarly, "empathy for employers, should not foreclose critiques of [and activist reactions to] individual and systemic in/actions that perpetuate the status quo for im/migrant workers" (Saxton and Stuesse 2018, 69). Market values harbor disabling, distancing, disturbing, and deadly results that afflict people throughout the life course, from preconception to the earlier-than-anticipated grave. Ultimately, folks can't work when they're injured, sick, replaced by machines, or dead.

Meanwhile, industry and state responses to farmworker health tend to emphasize farmworkers' individual responsibility for their health (Holmes 2013; Horton 2016b; Monaghan 2011; Salvatore et al. 2008) or outright dismiss their concerns. Many engaged anthropologists apply their insights to develop policy recommendations or interventions aimed at reducing harms in socially suffering and environmentally violated communities. Yet policies can be muted and weakened by corporate power on and off the farm. The "invisible harm" (Goldstein 2017) of ecological and social "toxic layering" (Swartz et al. 2018) is often externalized (Leigh, McCurdy, and Schenker 2001), legally sanctioned, or ignored. Market and ecosocial human values are not reconcilable. The long history of collaboration between the state and industry, some of which I witnessed during debates about the soil fumigant methyl iodide, raises concerns about the ability of policies alone to protect farmworker and environmental health.

3 · PESTICIDES AND FARMWORKER HEALTH
Toxic Layers, Invisible Harm

On a warm August day, fifty-two-year-old Mario, a farmworker and father of four children ages thirteen to twenty-five, rose before the sun came up to harvest raspberries under contract for a large grower-shipper company. Before the first break, around ten o'clock, he and other workers experienced dizziness and nausea. Some, including Mario, developed rashes. Mario felt his right arm go numb. It was difficult to move.

Later, during the half-hour lunch break, he approached his *mayordomo* (foreman) and informed him that he felt unwell. Other workers in the crew quietly shared their symptoms also. The *mayordomo* told them that they could keep working through the day and see how they felt, or they could go to the doctor's office for an exam.

Mario chose the latter option. His coworkers soldiered on with the harvest, afraid of losing a day's worth of work and income.[1] It did not feel like a real choice, even though this contradicted the company's own written policies and California occupational health and safety laws,[2] but causing trouble or failing to meet the harvest quotas could get them fired.

I first met Mario at a community gathering held at a school gymnasium. More than fifty farmworker families came to the event, which included free childcare and a meal of submarine sandwiches, chips, fruit, cookies, and bottled water. Activist, son of farmworkers, and recent college graduate Josué organized the event. He felt an urgent need to alert his fellow community members about the pending approval of methyl iodide, a highly toxic soil fumigant that was being considered as a drop-in replacement for yet another highly toxic soil pesticide, methyl bromide, which was in the process of being phased out of use in California agriculture and elsewhere in the United States. Many growers felt eager to find alternatives so they could stay in business, sustaining recent years of record-breaking strawberry production and profits.

Josué asked me to present some research I had been compiling about farmworker health, soil fumigants, and pesticide exposure to share at the assembly. I would need to not only translate my findings from English to Spanish but also ensure that the information—content, tone, and mode of delivery—was accessible and relatable to the audience. Many farmworkers are monolingual Spanish speakers, with limited literacy and levels of formal education, and some farmworkers from the states of Chiapas, Guerrero, and Oaxaca, for example, speak Indigenous languages almost exclusively at home (Arcury et al. 2006; Farquhar et al. 2008; Indigenous Farmworker Study 2010; NCFH 2018).

These skills, something I now routinely infuse into my teaching, community work, and engaged research practice, took time to develop. I arrived in the Pájaro and Salinas Valleys fresh from graduate coursework, where credibility and validation came from using words, theories, and frameworks largely meaningless to regular folks, including im/migrant farmworkers. I felt nervous and still unsure of what lay lingo farmworkers use to describe illnesses, disease, and pesticides. Josué graciously helped me interpret. Eventually I grew confident enough to deliver bilingual workshops on my own and to speak candidly with policy makers and lobbyists (see chapter 5).

After listening to the history of pesticide policies in California from a state assemblyperson and learning more about the global context of pesticide intensive agriculture from a staff member from the Pesticide Action Network, we left time for questions. Mario raised his hand. "What are some of the symptoms of pesticide exposure?" he asked. "And what should we do if we suspect we've been harmed by pesticides?"

I approached him afterward to learn more about what he had experienced. He also wanted to know about how he, as a farmworker, could get involved in antipesticide activism. A few weeks later, I visited Mario at his home in Watsonville. After he poured me a tall cup of orange-flavored Fanta soda, we talked about what had happened to him at work and in the months following his exposure. While he didn't remember seeing or smelling anything unusual that day, he suspected pesticides were responsible for his symptoms.[3] The numbness and heaviness in his right arm persisted for months, and the way his employer and the doctor he visited had handled his illness unsettled him.

During my visit Mario retrieved a copy of the Spanish-language employee manual that he and his coworkers had received during a mandatory safety training. He had highlighted in bright yellow a section stating that workers should report suspected pesticide exposure to supervisors and seek care immediately. With an eighth grade education, Mario could read and understand the manual. The same may not have been true for some of his coworkers.

The *mayordomo* had allowed Mario to leave and seek care. Before he could be seen by a doctor, however, a human resources staffer with the grower-shipper company interviewed Mario so that his complaint could be formally filed. When the staffer asked Mario to rank his pain and discomfort on a scale of 1 to 10, Mario indicated 7. The partial paralysis, tingling and numbness in his arm terrified him. He remembered watching the staffer circle the number 3, explicitly downgrading his condition and thus shaping the treatment he would later receive at the doctor's office. This parallels ethnographic findings that the pain and discomfort of women and people of color are routinely underestimated or dismissed by medical professionals (C. G. Castillo 2018; Holmes 2013) and industry (Quandt et al. 2000; Rao et al. 2004; Snipes, Cooper, and Shipp 2016).

On another occasion, I had been waiting in a grower-shipper firm's office lobby to interview someone in the workplace health and safety department. Behind the reception desk, I saw thousands of red file folders lined-up on some shelves, with more stacked around the floor. I casually asked the receptionist about them, and he informed me that they held years' worth of occupational injury, illness, and discrimination and harassment claims. It felt overwhelming and frustrating to think that cases like Mario's might end up here.

At the clinic, contracted via the company's workers' compensation insurance policy, a doctor treated Mario's rash with an over-the-counter allergy medication (loratadine) and a corticosteroid cream. Other farmworkers informed me that doctors routinely wrote off rashes and itchiness as mere *alergias* (allergies). No blood work or urine samples were taken, which might have revealed the presence of pesticide metabolites, chemical breakdown products or decreased enzyme levels that can indicate acute or chronic exposure (Arcury et al. 2009, 2014; Arcury and Quandt 2003; Bradman et al. 2007; Runkle et al. 2013). State officials did not investigate the incident at the farm where Mario worked. If they had, they would have interviewed the workers and maybe taken samples of their clothing or plants from the fields, which also can be tested for residues that exceed legal limits. Ultimately the doctor did not acknowledge or treat Mario's neurological complaints.

Not a single workers' compensation attorney would take Mario's case. The diagnoses and costs listed in the official injury report and the minimal medical care he received were not enough to attract a lawyer's attention. Workers' compensation lawyers are paid only if they win in court, which results in the privileging of cases that are most likely to result in major settlements (Saxton 2013). Settling in court also closes one's case permanently, preventing injured workers from accessing coverage if their work-related injuries or illnesses resurface or worsen, as Juan (see chapter 1) discovered.

As a final effort to see if we could figure out what Mario and his *compañeros/as* might have been exposed to that day, I contacted the Santa Cruz County Agricultural Commissioner's Office in Watsonville. Agricultural commissioners are appointed in each county by the deputy director of the California Department of Pesticide Regulation (CA DPR). They are responsible for enforcing pesticide laws; reviewing, issuing, and keeping track of pesticide use permits and applications; and investigating violations and exposure. I called the office beforehand to request the data and spoke with Eduardo, one of the staff members.

Mario and I arrived at the office, tucked away in a sprawling, concrete commercial building in Watsonville. There, Eduardo handed me a compact disk. He said it contained all of the pesticide use permits from 2010, the year of Mario's exposure, logged using Microsoft Access. Even though Eduardo exhibited sympathy for Mario and was trying to be helpful, the disk and the data it contained could only be accessed by those with a computer, the right software, and the ability to read and sift through endless lines on spreadsheets. This added another barrier to Mario's quest for information and closure. If nothing else, he at least wanted to know the chemical or chemicals he had been exposed to so that he could perhaps anticipate what to expect if his neurological issues continued.

Chapter 2 illustrated the ecologically and socially toxic products, processes, and social relationships in the strawberry industry. Pesticide drift—what Mario experienced—is pervasive and ubiquitous. It is the dust that clings to farmworkers' clothing and bodies after leaving the fields. It is also in the air and in the tap water in and around farmworkers' homes (Arcury and Quandt 2003; Bradman et al. 2009; Tracking California, n.d.; J. L. Harrison 2011; Rao et al. 2007; S. L. Steinberg and S. J. Steinberg 2008; S. J. Steinberg and S. L. Steinberg 2008). Pesticides travel far from their zones of application to other worksites, and to schools, daycare centers, hospitals, and homes. Exposure is not always perceptible. Many pesticides are colorless and odorless; they do not always cause an acute or immediately noticeable reaction in or on the body.

The quantities and pervasiveness of multiple pesticides and other industrial toxicants—from manufacturing residues to heavy metals throughout the food chain and in the environment, even far from their points of origin (Saxton et al. 2105)—can make it difficult to pinpoint the relationships between toxicants and disease outcomes, or to determine with certainty what or who is responsible for resulting "toxic epidemics" (Uesugi 2016). It is overwhelming to think about, but I believe that some of anthropology's tools can help mitigate panic and inform strategic reactions.

What follows in this chapter is my analysis of the many toxic layers perpetuating both visible and invisible harm endured by pesticide-exposed and -sickened farmworkers and their families. I am purposefully extending the definition of *toxic* to include pesticides *and* socially, economically, ecologically, and politically harmful conditions and relationships. I aim to rupture the distinctions between the visible and invisible, perceptible and imperceptible, actionable and nonactionable (M. Murphy 2006, 2008). Whether or not we can see, feel, smell, or understand environmental health and im/migrant farmworker problems depends a great deal on who is or isn't paying attention, what kinds of questions are being asked, and how toxic risks and exposure are framed in different policy, research, corporate, and community circles (Davies 2019). Critical medical anthropologists offer some concepts that I have found helpful in challenging the imperceptibility of exposure to toxic substances: *syndemics,* or how diseases and disparities intersect and exacerbate one another in bodies and communities, and *chronicities,* how diseases and disparities evolve over the life course in patterns that don't reflect the neat biomedical binary between acute (temporary) and chronic (permanent but perhaps manageable) conditions. Such engaged theoretical and conceptual models can assist us, as researchers and activists, in shifting approaches to toxic problems from singular chemicals and behavioral changes to more dynamic interventions, organizing, and reactions.

TOXIC LAYERING: UBIQUITY AND EMBODIMENT

Toxic layering is present in the quantities and scales of agricultural pesticide applications. In the years 2011–2012, global agricultural pesticide use reached almost six billion pounds. The U.S. tally reaches over a billion pounds per year (Atwood and Paisley-Jones 2017; L. Gross 2019). Of the 209 million pounds of pesticides applied in California in 2016, over forty million pounds were soil fumigants (CA DPR 2017b, 2017c).[4] Hundreds of different herbicides, insecticides, and fungicides applied directly on or near crops also account for a significant portion of agricultural pesticide use in California, the United States, and globally.

Each year the Environmental Working Group (EWG) produces its "Dirty Dozen" guide to help consumers mitigate their exposure to pesticides through the foods they eat (EWG Science Team 2020). Strawberries have consistently ranked among the highest for measurements of pesticide residues on purchased and washed fruit. They are among the top thirteen California crops that use the most pesticides, over 80 percent of which are soil fumigants (CA DPR 2015, 2016, 2017b, 2017c). Close to eleven million pounds of pesticides were used in California strawberry production in 2017 (CA DPR 2017d). The results for the EWG guide, based on U.S. Department of Agriculture (USDA) pesticide residue monitoring data on harvested crops in 2017, found eighty-one different pesticides on over a thousand samples of strawberries, most of which were grown in the United States (Environmental Working Group 2019).

These findings, while alarming, severely overlook farmworkers' exposure by emphasizing the *consumption* of contaminated fruits and vegetables.[5] Beyond residues on or in the foods we eat, farmworkers and their families endure heightened levels of occupational and residential exposure (Philpott 2011). For example, soil fumigants do not leave traces on or in fruits and vegetables, but they do pose direct hazards to farmworkers and fence-line communities (Guthman and Brown 2016a). Many of the most agriculturally productive regions of California border households, schools, and other farm worksites. This results in concentrated zones of toxicant exposure and embodiment in farmworker communities on California's Central Coast and inland in the Central and Imperial Valleys (L. Gross 2019; Philpott 2011; Prasad et al. 2017, 46–52; S. L. Steinberg and S. J. Steinberg 2008; S. J. Steinberg and S. L. Steinberg 2008).[6]

Official CA DPR reports of pesticide use are posted online each year, creating a virtual paper trail and giving a semblance of transparency about pesticide use rates and quantities that is hard to come by in other countries and U.S. states. Memos preceding the release of pesticide use data in California assert that the total amount of pesticides (measured as pounds per acre) applied in California has been decreasing (2018, 2017b, 2017c, 2017d). Similarly, a 2017 U.S. Environmental Protection Agency (U.S. EPA) report urges readers to not be overly alarmed by its findings on national pesticide use and sales trends, since "changes in the amount of pesticides used are not necessarily correlated with changes in the level of pest control or changes in the human health and environmental risks associated with pesticide use" (Atwood and Paisley-Jones 2017). These assurances are designed to encourage residents and consumers to keep eating their fruits and vegetables. They ignore region-specific pesticide use patterns that have population-specific consequences (Philpott 2011). Even if overall pesticide use has been trending downward, data from some of the highest-value crops, like strawberries and spinach, show more and multiple pesticide applications are happening per acre per year.

What do we know about how pesticides affect human bodies and our surrounding environments? What are the human health consequences of pesticide exposure? Many pesticides are designed to chemically disrupt vital biological functions, organ systems, cells, and DNA in pest organisms, ranging from fungi to insects. This may result in disablement, paralysis, respiratory issues, failed reproduction, and/or death, among other problems. For example, organophosphate pesticides, which are among the most commonly used in agriculture, work by disrupting cholinesterase, an enzyme produced in our bodies and the bodies of other organisms that is necessary for central nervous system and brain function. Along with hormones (from our endocrine system), enzymes and other bodily biochemicals enable our bodies to function, maintaining cardiac rhythms and respiration, brain messaging and cognition, gastrointestinal activities, reproductive processes, and development and growth in fetuses and children, to name but a few examples (Adeyinka and Pierre 2019; Bergman et al. 2013; Nicolopoulou-Stamati et al. 2017).[7]

Given these interconnections among ecosystems, organisms, bodies, and bodily systems and cells, many and multiple ecological and biological processes and functions can be affected by pesticide exposure at high or even low levels. Polyxeni Nicolopoulou-Stamati and colleagues note the complex variables: "The type of pesticide, the duration and route of exposure, and the individual health status . . . are determining factors in the possible health outcome. Within a human or animal body, pesticides may be metabolized, excreted, stored, or bioaccumulated in body fat. . . . Furthermore, high occupational, accidental, or intentional exposure to pesticides can result in hospitalization and death. (2016, 1, citing Alewu and Nosiri 2011; Mnif et al. 2011; Sanborn et al. 2007; World Health Organization 1990).

So many im/migrant farmworkers already endure multiple and chronic health problems—from diabetes, to heart disease, to stress and depression (Horton 2016b; Villarejo et al. 2010; McCurdy et al. 2015). That exposure to toxic substances could make them even more susceptible to disease and suffering should be even greater cause for concern and attention. The stories and layered illnesses and diseases of Berta and Joel, while not conclusively linked to pesticide exposure, explore a farmworker couple's lived experiences of syndemics and chronicities in an ecosocial framework.

BERTA AND JOEL: SYNDEMICS AND CHRONICITIES OF FARM WORK, IM/MIGRATION, AND EXPOSURE

I met Berta and Joel at a farmworker residential community on the outskirts of Watsonville. A graveled driveway tucked into the bend of a rural road led to Vista Valle Verde, a housing camp composed of about fifty families. Vista and other state- and county-subsidized farmworker housing units in the area had once been informal farm labor camps, often made up of barracks, old barns, and collections of wooden shacks owned and poorly tended by growers or landowners. Often these dwellings had no running water, or people shared a hose or tapped into irrigation systems to wash, bathe, or drink. Sometimes a single porta-potty would also be on-site, used by dozens of residents.

Joel and Berta had lived at Vista from the time when it was known as El Ollo (The Hole). They once showed me a few pictures from the 1970s. One featured Joel posing tall and strong in a doorway. A flap of tattered curtain was the only privacy they had. In the 1990s Monterey and Santa Cruz Counties tore down many of these decrepit dwellings and replaced them with modest single-family apartments, equipped with running water, private bathrooms, full electricity, kitchen appliances, heat, and septic systems. Berta and Joel were among the original residents to be rehoused following the renovations. At Vista sometimes the septic system overflows, and this is reminiscent of the old days when the nearby creek flooded, creating muck and stench. Shared resources at the modernized camp include a community room with a kitchen; some dated desktop computers that sit in a corner, unplugged, dusty, and without an internet connection; child-size desks and chairs; a playground and small soccer field; some space for gardening; assorted well-tended fruit trees; and laundry lines. Entire families, sometimes spanning three generations, occupy the apartments at Vista.

The camp is surrounded on all sides by lettuce, raspberry, and strawberry fields. When workers come home from the fields, or when they stay at home in retirement, they continue to face exposure and drift from toxic pesticides, along with their children who love to bike, skateboard, run, and play outside. Another camp that is privately managed sits across from a pesticide storage and mixing facility and is also surrounded by strawberries. On the road to Vista, I routinely witnessed (and smelled) pesticide applications. At one point, a pesticide applicator helicopter crashed in a nearby field across the road.

I first visited Vista with another *compañera* who, upon hearing of my interests, invited me to tag along with her to a Bible study group she attended at the camp. Berta, Joel, and others were hanging out at the central picnic tables, warming themselves in the sun, playing cards, gossiping, hanging laundry out to dry, tending to small garden plots of flowers and chiles, or watching over children, grandchildren, nieces, and nephews. The midafternoon crowd included former farmworkers and middle-aged workers who had been forced to retire early due to injury or ailment.

After about a month of my weekly visits and lots of listening, Berta and Joel would invite me to stay for meals. Others in the group asked if I could teach an adult English class once a week. In order to host them in the on-site community room, the organization managing Vista required me to sign a volunteer agreement; it expressly forbade me from interacting with residents outside official volunteer hours. I signed it, but I also made a judgment call: I would not turn down residents' invitations to get to know me better, to share tortillas and beans. At a different camp, a chain-link fence surrounded the property. Topped with barbed wire, it mimicked the enclosures around the medium-security prison and municipal garbage dump next door. The fences and the geographic isolation and hiddenness of Vista and other camps kept residents in, and largely (though not completely) succeeded in keeping unofficial visitors out, resulting in high levels of social isolation and despondency among the people I met.

The residents at Vista felt this acutely and in a binational way. Berta routinely wept for her recently deceased older brother, parents, and other siblings, some of whom died alone in México. She and Joel were never able to conceive children of their own. Diminished fertility for both men and women can be related to long-term pesticide exposure, even at low levels (Colborn and Carroll 2007; Mehrpour et al. 2014). Berta could not drive; Joel could, but his leg had been amputated and his vision continued to deteriorate, both consequences of untreated long-term diabetes. He had also recently suffered a stroke that affected his speech and mobility, on top of chronic back pain from several decades of stress and stooped repetitive labor in the strawberry fields. Joel limped his way around the camp with the help of a cane. Ultimately, the couple only left home for doctor's appointments at the county clinic and to buy groceries.

During one English class, I had folks draw pictures of their family trees to help them narrate and translate their kin relationships. This led to long conversations about life, disease, and death in México and the United States. Many of Berta and

Joel's siblings, also farmworkers, had passed away following heart attacks, strokes, and complications from diabetes long before their sixtieth birthdays.

During Christmastime 2010 the local food bank, the Salvation Army, and other charities came to Vista to distribute toys, clothes, and special holiday foods. Residents prepared a feast of Mexican dishes lined up on tables in the community room: pozole, enchiladas, colorful gelatins, crunchy tostadas topped with refried beans, mayonnaise-heavy salads, and vinegary *cueritos* (pickled pigs' ears). Kids ran around outside, screaming with glee, free from school for almost an entire month.

Even amid the festive atmosphere, most residents came to the community room, filled their plates and bowls, and went back to their individual apartments to dine alone. Berta invited me over to eat that day and seemed especially despondent. She appreciated the good food, but at the bottom of her food bank bag of canned goods she found a shoddily crafted and somewhat dirty hand-knitted hat. She tossed it to the side with disdain for its crude construction, ugly colors, and for what it represented: a life of secondhand goods. Berta wept. She missed her family and the celebrations in their hometown. Sometimes, life at Vista felt like prison (*es como una cárcel*).

I last heard from Berta and Joel around Christmastime 2013, the winter after I left Watsonville. Berta had been diagnosed with stage III breast cancer. Decades earlier she had endured uterine cancer and a hysterectomy. Joel reported to me that his doctors found inoperable tumors on his spine. I do not and cannot know for certain whether or not Berta and Joel's multilayered and evolving health issues had anything to do with pesticides, but their problems cannot entirely be written off as natural aging. Their lifetimes have been marked by exposure to toxic substances, as well as constant levels of trauma, grief, poverty, social exclusion and isolation, and hardship. And none of this ended after they retired from the strawberry fields.

The sum of the invisible harm and toxic layering engenders the syndemics and chronicities of farmworker health and illness. Depression and diabetes (Mendenhall 2012), frustration-induced chronic stress, rage, high blood pressure, and heart disease (Horton 2016b) are intensified and exacerbated by social and environmental suffering. Manderson and Warren describe this "entwining" of diseases and disparities as "recursive cascades" wherein "social conditions . . . not only render the individual vulnerable to developing chronic conditions but also contribute to the ongoing generation/production of further chronicities [of disease and suffering]" (2016, 480). Interventions that encourage farmworkers to wash their hands and separate their laundry may reduce their exposure to toxic pesticides to varying degrees, but they are inadequate in and of themselves when it comes to addressing the intersecting everyday circumstances faced by Berta, Joel, Mario, and others.

TOXIC LAYERS AND (IN)VISIBLE HARMS

In some pesticide safety and awareness trainings and educational materials, farmworkers are advised to avoid hugging their children immediately upon returning from work. Children's small and still developing bodies are especially vulnerable to toxic substances, yet the demanding hours required in agricultural jobs mean that farmworkers rarely get to see or spend time with their kids during the season. How do farmworkers weigh these multiple and intertwining harms and risks in their daily lives?

Medical anthropologist Donna Goldstein defines "invisible harm" as "the broad effects of increasing environmental toxicity and contamination in specific late capitalist" or highly industrial but poorly regulated contexts (2017, 321). "Toxic layering" (Swartz et al. 2018) identifies and articulates the simultaneous or sequential exposure to pesticides and other toxicants and social suffering in people's everyday lives that accumulates over the course of a lifetime (Arcury et al. 2014) and across generations (Colborn 2006; Colborn and Carroll 2007; Colborn, Dumanowski, and Myers 1997; Guillette et al. 1998). Toxic layers and invisible harm include health outcomes and environmental affects that we can see, feel, and measure with certainty: those that are perceptible with our senses *and* validated (albeit unevenly) by researchers, state regulators, and medical and legal authorities, at least when we are paying attention.

Medical doctor, im/migrant, mother, and activist Mona Hanna-Attisha (2018) describes how sick kids, the extraordinarily high lead levels in their bodies, and the decaying urban public water infrastructures of Flint, Michigan were made visible through a combination of science, medicine, and civic action. Still, these facts were rendered illegible and nonactionable by those with the power to prevent and mitigate the public environmental health crisis that ensued and continues to harm impoverished communities of color in Michigan and beyond. Geographer Thom Davies notes that toxic landscapes like Flint and Cancer Alley (an industrial corridor in Louisiana) are only invisible to those who aren't looking at, counting, or living through exposure to toxic substances (2019, 11–12) and the "slow violence" or gradual deterioration of human bodies and lives that ensues (Nixon 2011, 3–4). The legal and de facto neglect, ignorance, indifference, and inaction experienced by Mario following his exposure, and in the chronic isolation faced by Berta and Joel that worsened as they aged, are by-products of the toxic institutional cultures—including corporate and state behaviors, design, and infrastructures—that perpetuate the reproduction of harms in industrial agriculture and im/migrant communities.

Pesticides are one among many toxic layers endured and embodied over farmworkers' life courses. Punishing labor regimes and inadequate or substandard working conditions (Holmes 2013; Horton 2016b), poverty wages and geographic and social isolation (Bail et al. 2012; Benson 2008a, 2008b; García and

Gondolf 2004; Mitchell 1996), food insecurity (Brown and Getz 2011; Carney 2015; Held 2018; Minkoff-Zern 2019), bad housing conditions (Arcury, Jacobs, and Ruiz 2015; CIRS 2018), and a punitive and deadly im/migration system (Cartwright 2011; Holmes 2013; Horton 2016a, 2016b; Kline 2019; Unterberger 2018) are among the many others anthropologists have explored. Pesticide exposure adds to farmworkers' overall structural vulnerabilities, or the patterned suffering endured by people who are especially marginalized in our society (Quesada, Hart, and Bourgois 2011).

The resulting cycles of ill health and disease are not so easily reconciled by laws, biomedicine, self-care, or interventions on what public health practitioners problematically call risk behaviors. In the area of farmworker pesticide exposure, these include not washing hands at or after work, laundering contaminated work clothes together with nonwork clothes, not changing out of work shoes or clothing before returning home, hugging or playing with one's children upon returning from work before washing up, not wearing long sleeves and pants and other personal protective equipment like gloves and masks, and eating unwashed fruit while harvesting (Bradman et al. 2009; Cabrera and Leckie 2009; Monaghan 2011; Salvatore et al. 2009). These strategies may mitigate but do not eliminate the multilayered risks of pesticide exposure (Salvatore et al. 2008). Bigger structural changes and paradigm shifts in research; knowledge generation, dissemination, and application; agricultural production; and policy making and enforcement are urgently needed not just to assuage and mitigate but also to eliminate the disease burdens im/migrant farmworker communities in particular endure from toxic layering (Nicolopoulou-Stamati et al. 2016).

Toxic Disconnects and Social Distances

So-called lay and expert or state-based perceptions of and experiences with toxicants routinely conflict and are not represented or respected in the same ways (Brown 2007; Cartwright 2013; Scammell et al. 2009). The results of scientific inquiries or state responses to toxics don't always match up with Mario's and others' embodied and grounded knowledge, or "how [people] view the world through their bodies" (Wilde 2003, 170) and the environments where they live, work, and play (Davies 2019; Peña 2005). Cultural attitudes about pesticides and health—among agribusinesses and farmworkers, but also within farmworker communities and the general population—are also fraught with tensions and disagreements (Dowdall and Klotz 2014; Guthman 2017, 2018, 2019; Guthman and Brown 2016a, 2016b, 2017; Saxton 2105b). This further complicates efforts to react to toxic environmental injustices and to holistically acknowledge, treat, prevent, and understand accompanying health problems.

Thus, pesticide exposure and the multilayered health issues it engenders— from womb to tomb—are examples of contested illnesses: patterns of environmental suffering and disease that are ignored or overlooked by people in positions of authority, privilege, or power and sometimes even denied by members of afflicted communities (Brown 2007; Farmworker Justice 2013). In other cases, contestation produces injustices when one toxically exposed group receives more recognition, validation, treatment, and support than another. This is evident in the disparate political responses to toxically produced injuries and illnesses globally, including within farmworker and peasant communities (Bohme 2015b).

For example, some Vietnam War–era veterans have received compensation from the U.S. Veterans Administration to help treat and tend to the illnesses and diseases—from diabetes to Parkinson's disease—developed following exposure to Agent Orange, the defoliant pesticide used in chemical warfare. Exposed Vietnamese communities have received no reparations or support from the U.S. government to date even though they bear some of the highest burdens from residual contamination (Uesugi 2016). Meanwhile, the helicopters used to broadcast this chemical (which continues to harm future generations of Vietnamese) have been repurposed to apply pesticides in California agriculture. Similarly, the 1984 gas leak at a Union Carbide pesticide factory in Bhopal, India, left thousands of workers and nearby residents dead, injured, and chronically sick and neglected (Fortun 2001). People born long since 1984, much like the children in Vietnam, continue to endure cancers, stillbirths and miscarriages, debilitating birth defects, and chronic illness and disability; the toxicants from the gas leak, from the dumping that preceded it, and the seepage that continues have never been adequately cleaned up (Pundir and Jain 2019; Bhopal Medical Appeal, n.d., 2013, 2016).

The lingering and deadly residues of Agent Orange in Vietnam and the Dow Chemical (formerly Union Carbide) gas leak in Bhopal may seem like extreme examples to include in this discussion about California farmworkers. Yet each group endures ongoing and intergenerational health consequences that affect not only those directly maimed by toxics but also the families and communities who care for them and the children and grandchildren born decades later. These are the hidden syndemics and chronicities of exposure to toxic substances: long-term, ever evolving, and intersecting illness, suffering, exhaustion, disability, and death that exacerbate one another, especially in the context of im/migrants' everyday stresses and their exclusion from health care and other social supports (as described in chapters 1 and 4).

In California, pesticide exposure can be extreme, as is experienced by the farmworkers who notice acute symptoms immediately. A strange smell or a suspicious-looking mist may precede itchy eyes, runny noses, sore throats, headaches, nausea, dizziness, vomiting, metallic tastes, numb tongues and limbs, fainting, heart palpitations, profuse sweating, twitching, and/or loss of consciousness (Adeyinka and Pierre 2019; Robb and Baker 2019). Others may see or feel nothing at all and still be affected later on, gradually, as life goes on. Still others may also experience anxiety and disorientation, on top of other issues. As environmental historian Michelle Murphy (2006) notes, however, the more latent and long-term consequences of pesticide and exposure to other toxic substances are not always perceptible, and in many cases are actively "externalized" (Murphy 2008, 596). Murphy refers to exposure to toxic substances and relations produced by industrialization and capitalism as a "chemical regime of living," a consequence not only of "new . . . innovations, but the accumulated result of some two hundred years of industrialized production" (2008, 697–698). If we use this framing to think about the history and evolution of agriculture in the twentieth and twenty-first centuries, we can see how the consequences of pesticides in the lives of farmworkers are actively "externalized" and "distant" from "corporate accountability" (2008, 697–698).

In these ways, exposure doesn't just result from accidents or mistakes (what are sometimes called acute events)—that is, spills or documented drift that results in immediately noticeable symptoms (J. L. Harrison 2011). They are more often routine and subtle, occurring throughout the everyday rhythms of life and work; they are sights, smells and sensations rendered ordinary through the established protocols and processes of industrial agricultural production that normalize toxics as neutral, harmless, and necessary. The gas leak at Bhopal and the era when Agent Orange was broadcast over Vietnam have brought on recursive cascades of chronic and syndemic issues. These cases illustrate, too, how systemic imperceptibility of exposure to toxic substances is exacerbated when a group is consistently denied personhood, rights, or an empathetic audience. Im/migrant farmworkers in the United States share this in common with Vietnamese civilians and the residents of Bhopal.

The ubiquity of pesticides at variable levels throughout one's lifetime—in everyday life, at work, at home, and at school—makes the hard distinction between *acute* and *chronic* less useful or convincing when it comes to understanding and reacting to the enduring and uncertain embodied consequences of toxics. Medical anthropologists Lenore Manderson and Carolyn Smith-Morris challenge the hard and artificial binaries between the acute and the chronic. An initially acute exposure could make a farmworker more vulnerable during future exposure—something Mario feared. Acute exposure may build, with lower levels of exposure accumulating over a lifetime, and contribute to the development and exacerbation of chronic diseases, disabilities, or untimely deaths. The terms *syndemics* and *chronicities* capture these long-range embodied interactions as "inequalities of all kinds . . . are compounded by disease and the experiences of chronic illness and disability" (Manderson and Smith-Morris 2010, 15–16).

Toxic layering and the harms (both invisible and perceptible), and the syndemics, chronicities and cascades they engender, are also a product of problematic research framings and toxic policies that fail to act on, let alone collect, the information needed to protect public and environmental health. This is evident in California, where many agencies are charged with studying social and environmental problems and developing and enforcing laws in the interest of public and environmental health.

Toxic Research and Legal Infrastructures

Anthropologist Kim Fortun attributes the "uneven distributions of environmental risk" as a consequence of how information about health hazards from toxics are "filtered" by different legal, medical, and social authorities (2001, 8). In the United States, an example of this unevenness is reflected in many of the first pesticide regulations, which sought to protect farmers from fraudulent marketing claims and consumers from toxic residues on fruits and vegetables (Mart 2015; Szasz 2007; Whorton 1974). Meanwhile, protections for farmworkers have been much harder to achieve, let alone enforce and maintain in practice (Bohme 2015a; Getz, Brown, and Shreck 2008; Mart 2015).

Mario's experiences suggest that behavioral changes on the part of workers are not enough to safeguard health. Still, much research and applied work focuses on intervening at the level of the farmworker (Horton 2016b). What more could be done to support farmworker health? Even though it can be difficult, empirically, to link pesticides and other toxic layers directly to some of the cascades of illness and disease farmworkers endure, what do we know, and how do we know it? Who produced the knowledge, and what are their interests? And what prevents different kinds of knowledge from being mobilized more immediately, proactively, and protectively? Even when there is compelling evidence about the health harms and hazards of toxic pesticides, it is not always put to good use.

For one thing, farmworkers have not always been included in state labor laws. California has gradually become more inclusive. In 2016, California farmworkers gained the right to earn overtime pay, and in 2020, eligible undocumented and im/migrant youth under the age of twenty-six became covered under the state health insurance plan, Medi-Cal. At the same time, California is complicit, albeit not uniquely so, in supporting market-friendly laws that minimize environmental health and safety regulations for agribusinesses and industries. Sometimes loopholes in laws intended to be protective and to reduce harm actually mitigate the laws' effectiveness, as do conflicts between legal protections and market values and expectations (Horton 2016b).

In the area of occupational health and safety, the CA DPR requires additional risk and safety evaluation of pesticides that goes above and beyond federal standards. It oversees pesticide product labeling; regulates and monitors pesticide use and permitting; assigns county agricultural commissioners to oversee pesticide regulations regionally; sets and enforces environmental, applicator, and agricultural worker protections; and conducts research into pest management strategies, including those that are less toxic or nontoxic (CA DPR 2017a; Malloy et al. 2018). Since 1991 the CA DPR has maintained extensive databases of state-wide agricultural pesticide use, recording the kinds and amounts used per crop, county, acre, and farm.

In 1986, California voters passed Proposition 65, the Safe Drinking Water and Toxic Enforcement Act, which requires companies to notify workers, consumers, and/or residents about products or chemicals that "cause cancer, birth defects, or other reproductive harm" (OEHHA, n.d.-a). For people who live near fields where toxic soil fumigants and other pesticides are applied, applicators must post bilingual notifications at residences, alerting households of the pending presence of carcinogenic chemicals in their surroundings. At the federal level, between 2017 and 2018 the new U.S. EPA Agricultural Worker Protection Standard (WPS) went into effect. The WPS prohibits people under eighteen years of age from doing pesticide application or working in recently sprayed fields. Farmworkers also have rights to access information about the pesticides being used in their work environments. Pesticide applicators are supposed to stop spraying if they see people nearby (Udall 2018). Still, knowing that you are potentially or actively being exposed to toxic or carcinogenic chemicals does not in and of itself protect you. Ultimately, the Trump administration is working hard to roll-back these gradual gains.

California also requires doctors to report incidences of possible pesticide poisoning and exposure, yet many "health care providers often feel ill-equipped to recognize or manage pesticide exposure or pesticide-related illness" (Quackenbush, Hackley, and Dixon 2006, 3) or lack adequate training in toxicology and occupational health (Kelley et al. 2013; Wong et al. 2015). Some doctors contracted to evaluate workplace injuries or illnesses and workers' compensation cases face pressure from insurers and employers, or the overwhelming volume of patients, to sign off on cases and get workers back to work as soon as possible (C. G. Castillo 2018; Saxton 2013; Saxton and Stuesse 2018). Perhaps this is what informed Mario's doctor's approach to care. This conflicts with the time actually required to heal and/or rehabilitate from different kinds of injuries, exposure, and illnesses. At the same time, occupationally injured or sick workers may be desperate to go back to their jobs because they can't afford not to work.

It is possible to measure levels of pesticides in soil, air, water, and human and animal bodies with special sensors, monitoring stations, and/or biomonitoring. For farmworkers, biomonitoring has most often taken place, albeit to a limited extent, through federal, state, or university research projects. The Center for the Health Assessment of Mothers and Children of Salinas (CHAMACOS) is a longitudinal cohort study involving farmworker families and their children in the Salinas Valley. University of California–Berkeley researchers with CHAMACOS, (which in Mexican Spanish slang means little kids or, more colloquially, "brats") took blood, urine, breast milk, umbilical cord blood, and placental samples, as well as swabs and collections of residues and dusts from workers' clothing and homes, and sometimes from plants in the fields, to measure levels of pesticides in farm-workers' bodies and surroundings. They wanted to see if relationships between exposure and child and maternal population health outcomes could be mea-sured empirically.

Not all of the CHAMACOS findings are clear cut, but some results suggest a number of adverse health outcomes related to chronic, long-term and multi-chemical pesticide exposure. These include but are not limited to respiratory difficulties, fetal and child developmental delays, and learning disabilities for children with preconception and pre- and/or postnatal exposure (Eskenazi et al. 2010, 2014; Gaspar et al. 2015; Gemmill et al. 2013; Raanan et al. 2016). While the U.S. Occupational Safety and Health Administration requires that "employers [in nonagricultural industries] conduct medical monitoring of workers exposed to harmful substances, [the U.S.] EPA has no requirements for monitoring of workers exposed to pesticides" (APHA 2010). In California, routine biomoni-toring of pesticide applicators, but not farmworkers, has been mandated by the California Medical Supervision Program of the OEHHA since 1974 (OEHHA 2017). The Centers for Disease Control and Prevention's *National Report on Human Exposure to Environmental Chemicals* estimates that 90 percent of the general U.S. population "have pesticides or their byproducts in their bodies" (CDC 2019; Chiu et al. 2018; L. Gross 2019). This is likely even higher among farmworkers and resi-dents who live near sites of industrial agricultural production.[8]

Some pesticides and other toxicants may also act epigenetically, whereby environmental exposure, social stresses, and traumas alter genes at rates that exceed the pace of human evolution. While only researchers can "see" molecules as damaged DNA or pesticide metabolites, molecular changes that manifest as diseases later in life include many different kinds of cancer (Alavanja, Ward, and Reynolds 2007; Zham and Ward 1998), neurodegenerative diseases like Parkin-son's (Brown et al. 2006), and birth defects (Rappazzo et al. 2016). It can also include "developmental effects [that] cannot be seen at birth or even later in life" (Colborn 2006) but are burdensome for people. These encompass even more of the invisible harm of pesticide exposure resulting from toxic layering in bodies and communities. Such molecular effects on bodies, environments, and health may be passed on to future generations (Brehm and Flaws 2019; Huen et al. 2015; Thayer and Non 2016). These are major concerns for Central Coast teachers, teen students, and their farmworker parents (see chapter 5), motivating much of their antipesticide organizing.

Also disturbing are findings of pesticides and other toxicants, once assumed to be safe at certain levels, that harbor subtle but troubling endocrine health con-sequences at levels well below established safety thresholds (Vandenberg et al. 2012). Paracelsus's five-hundred-year-old assumption, now embraced by agrichem-ical and other industries, has been "the dose makes the poison." Conventional wisdom previously dictated that "the lack of documentation means that toxic potentials can be ignored" (Grandjean 2016, 1), but that is no longer accepted in environmental health science circles. Disruptions or damage to the endocrine system and hormonal signaling may be implicated in several serious health problems, including reproductive disorders (Brehm and Flaws 2019; Giulivo et al. 2016; Piazza and Urbanetz 2019), fetal and child development and growth, diabetes and metabolic syndromes like obesity (Cox et al. 2007; Darbre 2017; Howard 2019; Lee, Jacobs, and Lee 2018; Ruiz et al. 2018; Velmurugan et al. 2017, certain cancers (Alavanja, Ward, and Reynolds 2007), and brain and mental health problems (Laurienti et al. 2016; León-Olea et al. 2014; Moretto and Colosio 2011).[9] In biomedicine and popular consciousness, these diseases are more often attributed to individual failings or bad choices and health behaviors, especially among poor and/or nonwhite individuals (Gálvez 2018; Guthman 2011; Montoya 2011). Mounting evidence, and the stew of toxic chemicals in the environment and our bodies, suggest that many other variables are likely at play, especially for the most exposed and vulnerable members of society.

Thus, toxic layering and the invisible harm it engenders are multidimensional, intergenerational, and multisystemic, from cells and bodies to households and communities, multitiered political structures, and intersecting ecological systems. The billions of pounds of the thousands of different pesticides used in industrial agriculture, along with the plethora of toxicants from other industries, create a toxic stew in our environment (Saxton and Sanchez 2016) and in our bodies. The tight control that pesticide and agribusiness companies enact over pesticide knowledge and policy—for example, by claiming that their compounds are trade secrets that cannot be disclosed or by fighting efforts to more rigorously study pesticides' ill effects—further clouds and blinds our perceptions about product safety and consequences.

Toxic Cocktails, Toxic Denials

The U.S. EPA's reliance almost exclusively on toxicity data provided by pesticide companies themselves is troubling; it creates serious conflicts of interest and perpetuates risks to public and farmworker health. Imperfect or missing data enable companies who produce and use pesticides to avoid responsibility for causing harms. In addition, pesticide safety evaluation and market approval processes usually focus on only one product at a time. In practice, pesticides are often mixed or applied as "cocktails" to boost their effectiveness at targeting multiple pests at once. The effects of this mixing on human health and bodily systems are even less well understood, but they certainly amplify farmworkers' multilayered exposure. An understanding of basic chemistry would lead one to believe that chemical mixtures are not necessarily benign even if the individual ingredients have been carefully vetted for safety. A pesticide drift incident on June 18, 2019, at a vineyard outside of Dinuba in California's Central Valley, for example, involved two insecticides being applied at the same time at a stone fruit orchard neighboring an active farm worksite. Sixty workers were exposed and three were hospitalized (Rodriguez-Delgado 2019b). One worker, Mardonio Solario, experienced severe respiratory issues. The labor contractors, the growers, and the pesticide applicators all violated a number of different protective policies (Rodriguez-Delgado 2019a).

The CA DPR tracks pesticide exposure and drift incidents throughout the state with assistance from county agricultural commissioners. These records are not free of biases, erasures, or access issues, as Mario and I found out. Often farmworkers are hesitant to make a report. In a 2010 drift incident report from Monterey County logged into the CA DPR's Pesticide Illness Query (CalPIQ), investigators commented, perhaps with some frustration, that workers did not tell their supervisors about their symptoms, refused treatment, or were not willing to be formally interviewed. Even though exposed workers may be suffering a great deal, from acute symptoms and from the fear that comes with suddenly being afflicted with painful, disorienting, and unfamiliar sensations, staying silent may feel safer, especially if one is undocumented (Arcury et al. 2006; Villarejo et al 2010). In other instances, workers who do get care are taken to occupational health clinics affiliated with grower-shipper companies, as was the case for Mario (in this chapter). Treating farmworkers in-house further distorts the data we have about pesticide exposure and drift incidents, as well as other work-related injuries, illnesses, and deaths (Horton 2016b).

At a Kern County field in 2017, ninety-two workers fell ill following a drift incident (Goldberg 2017). That same year, near Watsonville, over a dozen workers and their foreman got sick after a pesticide applicator sprayed four different pesticides at the same time on a neighboring field and didn't follow the legally required notification protocols (Goldberg 2018b). Instead, knowing that farmworkers were nearby but likely facing pressure to get the job done, the applicator told the foreman to notify him if anyone felt sick and he would stop spraying.

A pesticide drift incident in Monterey County in 2017 involved farmworkers transplanting celery starts. Seventeen workers experienced headaches, nausea, numbness in their extremities, impaired vision, respiratory problems, dizziness, and vomiting. The company haphazardly transported most of them to the hospital in private vehicles, though some workers drove themselves. After the Monterey County agricultural commissioner and the CA DPR completed their investigations, the agribusiness received a five-thousand-dollar fine (Goldberg 2018a). First responders should have been called to do a more thorough on-site evaluation of the workers' health, to take appropriate decontamination measures, and to ensure safe transit of people reporting illness to the hospital. The company boasted that most of the sickened and exposed employees returned to work the following day. Asserting this further deflected the company's accountability, and diminished sustained attention from investigators, journalists, and the general public. Formal inquiries into drift incidences like the ones described here usually only happen if a pesticide exposure is obvious and extreme (J. L. Harrison 2011). The company implicated in the Monterey County drift incident denied outright that pesticides had anything to do with the farmworkers' illnesses, despite strong evidence that they had been exposed to nine different chemicals (Goldberg 2018a).

Reported drift and exposure incidences recorded in the CalPIQ database also highlight the uncertainties and doubts of incident investigators, who are charged with being objective and impartial. Workers' accountings of their symptoms are often listed as "possible" or "probable" results of exposure and rarely "definite," even when multiple workers or residents convey a shared experience. In other instances, there is a failure to consider or collect all available data. A 2010 drift incident in a Monterey County strawberry field exposed at least forty people (farmworkers, supervisors, and applicators) to the miticide fenpyroximate, which is listed as "moderately hazardous" by the U.S. EPA and the World Health Organization (Pesticide Action Network, n.d.-a). Additional tests were ordered on some foliage samples, but the test for the chemical in question was not ordered and the packing house's preoccupation with getting the berries to market limited the information that could be gathered.

Farmworkers themselves don't always understand or acknowledge the relationships between their health and working and environmental conditions. They may think their symptoms are related to colds or are the result of an inherent physical (Snipes et al. 2009) or spiritual (Baer and Penzell 1993) weakness. Growers sometimes assume their workers are just hungover (Benson 2008a). Farmworkers who talk about their symptoms or request medical treatment are often accused of trying to get out of work early or even feigning illness to take advantage of workers' compensation insurance (Holmes 2013; Horton 2016b). It is also not uncommon for employers to avoid calling emergency responders when workers get hurt, fall ill, or lose consciousness, as Honesto Silva Ibarra (see chapter 2), Mario (in this chapter), Aniceta (see chapter 4), and the workers of the drift incidents all reported. The fact that toxically afflicted communities are often multiply marginalized and exposed, and their observations and embodiment of toxicants written off by people in positions of authority and power, further obfuscates the connections between exposure and ill health and disease (Bohme 2015b; Brown 2007; V. Das 1997; J. L. Harrison 2011; M. Murphy 2008; Uesugi 2016, 465).

Toxic Uncertainties

Toxic layering also manifests in the perpetuation of toxic ignorance, or a failure to appreciate unforeseen consequences of pesticide exposure (Morello-Frosch et al. 2009). What we don't know can and does hurt us. Even with significant environmental justice victories that have strengthened regulations and banned some of the most toxic pesticides, there remain gaps, oversights, and contradictions that leave some groups more vulnerable than others. It is notable that throughout the over one-hundred-year history of pesticide regulation in the United States, many of the laws passed have prioritized concerns about hazards posed to consumers through residues on purchased products, and growers' rights to have access to effective pesticide products, but not worker safety (Mart 2015; Pulido 1996; Whorton 1974).

The environmental researcher and writer Rachel Carson's 1962 book *Silent Spring* documented the effects of pesticides in the U.S. ecosystem. It challenged popular assumptions about pesticides as a marker of status and progress, technological advancement, economic growth, and "better living through chemistry." Her work is often credited with instigating the development of federal environmental policies, including the founding of the U.S. EPA. Ultimately Carson's public scholarship contributed to the U.S. ban of organochloride pesticides like dichlorodiphenyltrichloroethane (DDT). This spurred the development, marketing, and expanded use of organophosphate pesticides, which are less acutely toxic but harbor more dangerous and deadly long-term effects for farmworkers through routine exposure (Mart 2015). Environmentalists largely ignored farmworkers, who themselves were organizing in the 1960s and 1970s for the right to safe and healthy workplaces and communities (Mart 2015; Peña 2005, 2011; Pulido 1996). Pesticides banned in the United States often end up being dumped and marketed more aggressively in third world countries, where they harm other agricultural workers. This is what David Weir and Mark Schapiro term "the pesticide treadmill" (1987). The ubiquity of toxics is facilitated by a lack of global governance and consensus and significant power imbalances between foreign corporate agribusinesses and the countries where they produce food for export.

Toxic ubiquity and uncertainty contribute to fraught interpretations of how data and results are applied in occupational and environmental health policies and industry practices and may also lead to reactions of helplessness, hopelessness, and indifference. In the United States, protective or precautionary policies are rarely passed unless adverse health outcomes can be proven with certainty (Cranor 2017; Grossman 2014). Lingering uncertainty may lead researchers and regulators to call for more yet research or to remain apolitical when it comes to recommending how their results be interpreted or applied.

In 2011, following pressure from a civil lawsuit, *Angelita C. vs. California Department of Pesticide Regulation*, that claimed that disproportionate pesticide exposure for California Latinx schoolchildren constituted racial discrimination, the CA DPR and the U.S. EPA placed air quality monitoring devices at school sites abutting fields (Buford 2015; D. Jones 2011). While the EPA "issued a finding of racial discrimination" (Californians for Pesticide Reform 2014), this only resulted in a decision to do more testing around sensitive sites like public schools. Being overly conservative about the broader significance and implications of research findings can also perpetuate environmental violence through political ineptitude

FIGURE 13. The names of pesticide companies, as seen in San Quintín, Baja California, México. (Photo by the author.)

and inaction. In 2014, two years after official monitoring began, the CA DPR and U.S. EPA announced that the levels of pesticides in the air around rural California schools were well below the state-mandated thresholds (Californians for Pesticide Reform 2014). Ultimately, none of the people who filed the original civil suit received any compensation or acknowledgment that their concerns were valid (Buford 2015). The very public announcement of the results from the CA DPR and CA EPA intended to assuage the public and deflect attention away from a still ongoing problem.

The toxic methodologies used by pesticide companies to test for health hazards rarely take into consideration the variable levels of exposure that occur in and around farm fields. They also do not assess what happens when people are exposed to more than one pesticide at once or at variable or low levels over the life course (Froines et al. 2013; Mie, Rudén, and Grandjean 2018). How and at what levels pesticides in the air and water are deemed "acceptable" or "safe" in the first place are the results of problematic assumptions and imprecise selections of data that are not immune to market biases, research design flaws, or institutional neglect. Notably, in late 2010, during then governor Arnold Schwarzenegger's tenure, former CA DPR deputy director Mary-Ann Warmerdam approved the soil fumigant methyl iodide for use in California agriculture, despite widespread concern and protests from the scientific research community, the general public, farmworkers, environmental justice groups, and the CA DPR's own Scientific Review Committee (Froines et al. 2010, 2013). Methyl iodide first received approval from the

U.S. EPA in 2007, despite serious concerns about the pesticide's known carcino-genic, endocrine disrupting, neurotoxic, and reproductive effects raised by promi-nent scientists, including Nobel Prize–winning chemists (Cone 2007; Philpott 2007). During methyl iodide's registration, the U.S. EPA employed a number of officials with industry ties to and former employment in pesticide companies (Philpott 2007), including the fumigant's manufacturer, Arysta LifeScience. Warmerdam served as the deputy director of the CA DPR from 2004 to 2011 before resigning and moving on to work for Clorox (Huber 2011).

The people chosen by elected state governors and U.S. presidents to lead envi-ronmental, health, and occupational safety agencies are not always disinterested actors. Many have past and present ties to agribusinesses or other polluting indus-tries, and when their terms are up, they often find jobs elsewhere in the industrial sector. The "revolving door" effect of employees who rotate jobs between indus-try and public office or service, and the circulation of funding from private industry into the research programs of public universities, is commonplace (Ores-kis and Conway 2010). The Poison Papers Project, which, compiles tens of thou-sands of documents dating back as far as the 1920s, shows long-term cooperation between industry and the state to hide information about the known harms of pes-ticides and other industrially manufactured chemicals (Poison Papers, n.d.). This brings a number of ethical concerns about the ability of the law to protect (and our assumptions that it will), as well as exacerbating public distrust in state and elected officials (Lantham 2019; Meghani and Kuzma 2011).

Meanwhile, the United States has increasingly retracted support to fund sci-ence, research, and development in the public interest (Nader 1999). In public land grant universities with large agriculture programs, researchers whose findings con-trast or conflict with agribusiness or pesticide company funders' goals often meet intense hostility, including job loss and threats to their lives (Aviv 2014; Oreskis and Conway 2010; Van den Bosch 1989). Such was the case for African American biologist Tyrone Hayes of the University of California–Berkeley, who conducted extensive research on the endocrine disrupting and reproductive effects of the herbicide atrazine on amphibians. Syngenta, the company that manufactures and markets atrazine, funded Hayes's early work before cutting ties and engaging in a campaign to discredit his reputation when his findings showed serious develop-mental irregularities in exposed frogs that, if made public, would interfere with product sales (Aviv 2014). In contrast, University of California–Riverside profes-sor emeritus of plant pathology, Jim Sims, who spent his career researching the pest control properties of methyl iodide, vehemently defended the state of Califor-nia's decision to register it in 2010 (Warnert 2010b).

Even when state, university, or other scientists find and share concerning and convincing evidence about the harms of pesticides, it is not always acted upon by those in power (Poison Papers 2019). The U.S. EPA under the administration of President Donald Trump has ignored the cautions of its own staff researchers and actively rolled back or refused to enact protective policies following findings that the organophosphate pesticide chlorpyrifos harms children's brains and development. The U.S. EPA followed suit when scientists provided evidence that the popular herbicide Roundup (glyphosate) is a known carcinogen (Dennis 2017; Eilperin and Dennis 2017; Holden 2019). The efficacy of different individual pesticides is heavily studied and documented in field trials and lab studies run by pesticide companies, university cooperative extension agencies, and state and federal governments. In contrast, the long-term health and ecological effects of these pesticide mixtures remain underexplored or outright denied (Froines et al. 2013).

When concerning studies come out, companies may also hire private research firms to release contradictory data. This was the case when CHAMACOS researchers published results about the relationships between farmworker mothers' residential proximity to methyl bromide applications at different stages of their pregnancies. They observed potentially adverse fetal growth and birth outcomes (Gemmill et al. 2013). In response, grower-shipper companies commissioned researchers at the private firm Exponent, whose scientists produced a report aiming to delegitimize, downplay, and discount the evidence (Alexander 2013). Their findings supported claims that pesticides, and especially soil fumigants, could be used safely when applied in compliance with existing laws and product label directions.

All of these issues and lingering uncertainties raise a number of concerns about the quality of pesticide governance, even in a state like California, which is often branded as progressive when it comes to its more thorough pesticide evaluation processes and environmental and occupational health and safety regulations. Our ability to address knowledge gaps and uncertainties cannot keep up with the rates at which new toxics are introduced into our homes, workplaces, food and water supplies, and surrounding environments (Cranor 2017). Meanwhile, some states, counties, and communities have taken a more precautionary approach to pesticides.

Toxic Dependencies and Markets

As environmental sociologist Jill Lindsey Harrison observed during her research with environmental justice and farmworker activists in California's Central Valley, the culture of doubt about pesticide hazards is embedded within the very culture of agriculture: "Researchers who have studied farmers' perceptions of pesticide risks have found that farmers widely believe that pesticides are safe and controllable, spills during mixing and application are the only significant sources of pesticide exposure on farms, and pesticide residues do not pose a health hazard. This widespread causal disregard for the dangers of pesticides justifies the high rates of pesticide use and risky application practices" (2011, 72). Along these lines, many California growers are averse to what they see as excessive regulatory oversight. They complain of the challenges of managing farms and having to fill out copious amounts of pesticide use permissions paperwork, which must be reviewed and approved by state or county officials. This is especially true for pesticides deemed potentially dangerous enough to merit "restricted use." Such restrictions can include rules about the time of day and climate conditions during applications, and prohibitions around reentering fields hours, days, or weeks after applications. The creation of buffer zones around the edges of fields, ranging from tens of yards to a quarter of a mile—places where pesticides cannot be applied and/or that limit the quantities of a given pesticide that can be used at one time—are also intensely resisted by agribusinesses, as this takes high-rent land out of production.

Growers, grower-shipper organizations, and agribusiness companies often assert themselves as responsible stewards of land and natural resources and caring towards their workers. They routinely insist that voluntary pesticide safety provisions would be sufficient to protect environmental and worker health (J. L. Harrison 2011). They may frame environmental justice groups, farmworkers, and other concerned members of the general public as ill informed, irrationally fearful, and uneducated about pesticides and what it takes to grow food, while at the same time, underestimating and denying risks with well-funded public relations campaigns and responses (Hawkes and Stiles 1986; Rao et al. 2004; Saxton 2013).

These attitudes were evident at agricultural health and safety and pesticide conferences I attended, including one sponsored by AgSafe, a nonprofit organization that helped growers, pesticide applicators, and farm labor contractors and managers keep up to date about state and federal occupational and environmental health and safety policies and labor laws. Researchers and industry representatives from universities, legal firms, insurance, farm equipment, and pesticide companies mostly discussed procedures and products for preventing workplace illnesses and injuries. At a panel on managing emergencies in the fields, one industry representative, who worked for a large grower-shipper firm as the director of safety and operations, spent the hour mocking farmworkers, community members, emergency responders, and the media for their reactions to a pesticide drift incident that occurred in some carrot fields in Kern County. He resented how the portrayal of the incident on the news, including alarming images of first responders suited up in full hazmat gear, made the industry look culpable and irresponsible. Instead of reviewing protocols for hazardous incidents, he emphasized different public relations strategies to manage a company or employer's reputation and to mitigate panic from workers and neighbors.

Methyl Bromide Alternatives Outreach (MBAO) hosts a convention every year at which agricultural and food trade and processing industries affected by the United Nations Montreal Protocol mandated phasing out of the soil fumigant share the latest research on the prospects of farming profitably without this pesticide. I attended an MBAO convention in San Diego in November 2013 as a participant observer. Many presenters shared exciting research and efforts to replace methyl bromide and other soil fumigants with more integrated pest management (IPM) methods. IPM involves biological or cultural pest controls, such as crop rotation, introduction of predator species, grafting onto pest-resistant rootstock, or soil solarization and amendments to prevent or curb pest damage. Toxic pesticides are used only as a last resort and in moderation after all nontoxic and less hazardous methods have failed.

Still, the cultural attachment to agrichemicals and toxic soil fumigants and trust in pesticides as safe if used as directed remains a strong part of the culture of industrial agriculture. An extension researcher who gave a talk about her work comparing different combinations of still legal soil fumigants like chloropicrin and 1-3-dichloropropene (better known as 1-3-D, and by the trade name Telone) referred to the latter as her favorite fumigant, the "gold standard" for soil fumigation. At another panel I had a conversation with a professional pesticide applicator who worked for Tri-Cal, one of the largest soil fumigation firms founded in California in the 1960s, with operations expanding nationally and internationally (TriCal, Inc., n.d.). He insisted that if the individual components of fumigant "cocktails" were deemed safe by the state, then the mixtures and the so-called inert ingredients they are mixed with are too.

I felt skeptical upon hearing this. Evidence left behind by past and current generations of soil fumigants and other pesticides should be cause for grave concern. The chemical 1,2,3-trichloropropane (1,2,3-TCP) is a nonactive ingredient that ended up in supplies of the soil fumigants 1-3-D and dichloropropane-dichloropropene during production from the 1940s to the 1990s. Rather than remove it from their products, Dow Chemical and Shell decided to leave 1,2,3-TCP in their fumigant formulas and list it as an active ingredient even though they knew that it did nothing for pest control and even as concerns had been raised about its carcinogenicity. Since then, communities throughout California, especially in farming regions, have found levels of 1,2,3-TCP in water supplies that are significantly higher than those deemed safe by state health officials (Clean Water Action 2016; Khokha 2017; Thomas 2017). A $22 million lawsuit filed by the city of Clovis, just northeast of Fresno in the Central Valley, seeks retributions for exposure victims who drank contaminated city tap water (A. Castillo 2016). While the success of this case is encouraging for other rural California communities that will likely file their own lawsuits, it is also certain that reparations in the tens of millions of dollars will not adequately cover all the known and unknown health costs, current and future, associated with pervasive exposure to 1,2,3-TCP. Furthermore, Dow Chemical and Shell, the companies that marketed 1,2,3-TCP most heavily in California, are working hard to avoid responsibility for funding remediation methods that would lower levels in public drinking water to the "California Office of Environmental Health Hazard Assessment . . . public health goal ... of 0.0007 [parts per billion]." Even this number is far below the current State Water Resources Control Board level of concern, at 0.005 parts per billion (Thomas 2017). Agencies within the same state government make determinations about "safe levels" that conflict with one another put public health at additional risk.

Even with the documented, probable, and suspected risks and hazards of pesticide exposure for human and environmental health, agribusinesses and many growers vehemently defend their use of pesticides. A major argument is that pesticides, often referred to as "crop protection tools" to ease public uneasiness, contribute to public health by boosting yields of fresh fruits and vegetables that enable people to enjoy healthy lifestyles. The relationship between pesticides and increased yields, it is posited, also enables im/migrant farmworker well-being via job creation.

During a historic February 2012 Monterey County Board of Supervisors meeting, where growers, trade groups, environmental justice and farmworker activists debated the wisdom of the state's approval of methyl iodide, an agribusiness lawyer insisted that "without an effective alternative to methyl bromide, unfortunately the sky will fall on agriculture. The strawberry and lettuce industry which [are] dependent upon high yields to compete, will shrink. Capital will go and is going elsewhere. Jobs will be lost ... headquarters will not be built." Absent in this pro–methyl iodide argument is evidence that even with intensive and less-restricted use of soil fumigants from the 1960s through the 1980s, U.S. strawberry firms had already been expanding their production globally. A lot of different factors contribute to these moves. California agribusinesses' over-one-hundred-year-long asserted entitlements to land and water are under threat due to climate change and drought-induced water scarcity, contamination, and soil erosion and subsidence. The exponentially rising costs of fuel and owning or renting prime farming lands and processing plants, especially on the Central Coast, is another factor. Continued and heightened political assaults on the largely im/migrant farm labor force, the systemic cultural and economic devaluation of farm labor, and the struggles farmworkers face in attempting to sustain themselves and their families on poverty wages in communities with ever rising housing costs (see chapter 1) are other variables. And the allure of cheaper land, cheaper labor, cheaper water, and fewer labor and environmental regulations in other countries are yet others. All are implicated in the movement and expansion of California agriculture abroad.

In 2017 Dole, a once significant grower-shipper of California strawberries, announced its plans to close down its planting and production operations in California. Hundreds of farm and packing plant workers and truck drivers were laid off following Dole's transition to a publicly traded company. This coincided with the expansion of Dole's export-oriented berry farms south of the border, from Argentina to México (Ibarra 2017; Karst 2018; Mohan 2017). Dole's abrupt departure from the Central Coast is not the first time something like this has happened. The Pájaro and Salinas Valleys used to host major canning and produce freezing and packing industries, employing significant numbers of Mexican im/migrants and Chicano/a residents. Most of these plants shut down in the 1980s following striking workers' demands for better pay and working conditions (Bardacke 1994; Zavella 2011) and the global reorganization of food production and processing. Freezing companies sought cheaper rents and labor costs in places like Irapuato, Guanajuato, México, and took advantage of more lenient labor and environmental regulations that only grew more permissive with the North American Free Trade Agreement (Borrego and Zavella 1999; Zavella 2001). Major berry grower-shipper companies like Driscoll's, Giant, Naturipe, and Well-Pict are thus far maintaining headquarters in Salinas and Watsonville at the same time that they are scaling-up growing, packing, marketing and sales operations internationally at exponential rates. Driscoll's alone has growing and marketing operations on six continents, from México to Morocco.

Chicana anthropologist Patricia Zavella demonstrates how transnational migration between California and México involves complex labors of social reproduction and agricultural production. The creation of a class of disposable workers in México, who migrated to work in the factories and fields of California, found themselves doubly displaced when the freezers and canneries moved back to México in the 1980s. Chronic unemployment and underemployment on both sides of the border, and cycles of political and economic unrest in México, benefit the strawberry and other agricultural industries but severely limit and curtail the lives and life opportunities of im/migrant workers, and particularly for women (Zavella 2001, 2011). Even statewide expansion of the strawberry industry and its promises of economic growth for struggling communities and good jobs—for the farmworkers who harvest and pack, the truck drivers who distribute, or even recent college graduates, some of whom are second generation Americans seeking indoor office jobs or outdoor agricultural science, technology, engineering, and mathematics jobs with stable salaries and benefits—host formidable uncertainties and precarities. Compounding this, then and now, are anti-im/migrant policies that are triply and quadruply displacing workers and upending mixed-status families in agricultural communities.

At the same February 2012 Monterey County Board of Supervisors meeting, a leader from the California Strawberry Commission trade group aligned worker welfare to the health of strawberry fields as achieved via soil fumigation:

> "The strawberry in the strawberry [*starts to say "industry," but changes language*] *community* generates over seventy thousand direct . . . on-farm jobs in California. In Monterey County, the strawberry farmers bring nearly one billion dollars to the local economy. We have to have clean plants, healthy plants to grow strawberries. There's no choice. . . . [Farmworkers] . . . optimize every day in the calendar

to make money, to bring home money so we can pay our rent, our bills. Every day is crucial during the season. And when you have a field that is infected with some soil disease, the plants are wilting, very limited amount[s] of strawberries are on the plant, it's an obvious choice. [*Chuckles.*] Many of the workers will just come to me and say, "We can't make it here!" So clean soil is integral to the farmer, of course. But the workers are being forgotten here, they have to make a living. And each year there are a certain amount of days they can do that in. They optimize the days.

Missing from this argument is acknowledgment, on top of the invisible harm of pesticide exposure, that the grueling pace the farmworkers must maintain in order to make the piece rates and quotas that pay their rent and bills is also maiming to farm and food workers (Horton 2016b; Saxton and Stuesse 2018; Smith-Nonini 2011; Stuesse 2016).

RESISTING TOXIC CULTURES

Anthropological research has contributed to critiques about pesticides and health. These include documentations of accidental child poisonings (Swartz et al. 2018), the suicides of farmers in India indebted to intertwined pesticide, fertilizer, and seed corporations (Aggarwal 2008; A. Das 2011), and the syndemics and chronicities of ongoing agrichemical disasters (V. Das 1997; Fortun 2001; Uesugi 2016). Interdisciplinary medical anthropologists are also exploring the longitudinal molecular-level harms of exposure embodied in farmworkers as they age (Quandt et al. 2017). By stepping out of my disciplinary comfort zone to learn more about pesticides and health, I enhanced my ethnographic understandings of the vulnerabilities of the human body to toxics. This also supported different kinds of antipesticide activism, as I will describe in chapter 5.

Despite a disappointing outcome with the 1999 civil lawsuit against the CA EPA and CA DPR, teachers, students, health care workers, and farmworker activists in the Pájaro and Salinas Valleys continue to advocate for environmental health and justice and better protections. One farmer has even transitioned a field near Ohlone Elementary School in northern Monterey County to organic strawberry production and invites students and teachers over for educational tours (Cusick, n.d.). In 2019 the newly elected California governor Gavin Newsome issued a ban on chlorpyrifos (Dennis and Eilperin 2019). Some cities and school districts in California, including Watsonville, are restricting the use of the carcinogenic herbicide glyphosate (Roundup) on city, public, and school lands (Ibarra 2019). Several European Union countries, while far from perfect, more often follow the "precautionary principle" when it comes to regulating the use and sales of pesticides and other toxic chemicals. This urges that "protective action . . . be taken" when there is "substantial, credible evidence of danger to human or environmental health . . . despite continuing scientific uncertainty" (Grossman 2014). Meanwhile, in the United States, the burden of proof that a chemical is harmful is put upon those most affected by pesticides, who possess the fewest resources to adequately navigate regulatory bureaucracies and challenge built-in inequalities and hazards (Grossman 2014; J. L. Harrison 2011).

Engaged and activist anthropologists often make policy suggestions aimed at protecting or reducing harm for structurally vulnerable communities. Laura Nader, in her writings on contemporary legal anthropology (2002, 77), places a lot of stock in the power of the law to achieve social justice. Strong policies require enforcement, effective systems of accountability, broad inclusion of different workers and workplaces, and a culture of precaution even when we don't know everything. Pesticide companies are quick to circulate findings that negate even the most well-documented harms (Oreskes and Conway 2010; Poison Papers 2019). In stating this, I am not arguing for fewer regulations or restricted governance or accountability for agribusinesses and other extractive, polluting, harming, and maiming industries. Instead I want to reiterate what Jonathan Lantham (2019), executive director of the Bioscience Research Project, has cautioned: that single substance approaches to policy making cannot protect us, given the rapid rate at which new toxics are put on the market and released into the environment. Observing that many studies on farmworker health focus narrowly on workplace conditions, Elizabeth Cartwright urges us not to forget about "the larger toxic legal and social webs that define the quality of their lives" (2011, 480; see also Saxton and Stuesse 2018).

Ecosocial, layered, and invisible problems require fractal, dynamic, multipronged, and long-term responses and involvements. The frameworks of toxic layering and invisible harm help nuance our understandings of and activist responses to the lived experiences of pesticide exposure, and syndemics and chronicities challenge the problematic time- and space-bound framings of exposure to toxic substances as moments in time or bounded to specific places, bodies, or generations. We should maintain a healthy skepticism about market-based praise of pesticides and institutional neglect or ignorance around risks—those both known and as yet unknown. We must also become more chronic and syndemic, meaning multimodal, longitudinal, and persistent, in our research and activist practices to better attend to these challenges instead as opposed to the piece-rate approach to antitoxic activism. Accompaniment may provide a model for moving onward, and that is the subject of chapter 4.

4 · ACCOMPANYING FARMWORKERS

On a hot Sunday in August, I traveled south on Highway 101 toward Greenfield, a small agricultural city in the southern Salinas Valley. Every year Greenfield hosts El Día del Trabajador Agrícola (Day of the Farmworker) to honor the community's contributions to California agriculture and celebrate the diversity of the Indigenous Oaxacan im/migrant communities who have made the city their second home.

Sundays in the Pájaro and Salinas Valleys interrupt the work grind, *el jale*, in preparation for yet another week. Farmworker families attend church, sleep in, go grocery shopping or visit the *remates* (flea markets) for fresh produce, second-hand clothes, tools, toys, and odds and ends. Many folks wash their clothes at neighborhood *lavanderías* (laundromats), watch or play *fútbol*, and visit parks and playgrounds. In Watsonville, older retired farmworkers hang out on the benches or perch in wheelchairs and walker seats at the plaza, playing cards, sharing stories and *chisme* (gossip), and people watching.

On this August day, and in anticipation of the Día del Trabajador Agrícola festivities, Greenfield residents gathered at Patriot Park. Men donned freshly pressed jeans and embroidered button-down shirts, cowboy hats, and finely crafted leather boots. Some of the women wore long, red, handwoven *huipiles* (embroidered blouses and shirts), the customary *traje* (garb) of the Triqui Indigenous community. Mothers pushed babies in strollers as older children followed behind.

A mix of regional music, popular in different parts of México—*norteño, banda, cumbia,* and *chilena*—blasted from speakers. Few people danced other than the *baile folklórico* (Mexican folk dancing) groups providing programmatic interludes. Food vendors sold everything from energy drinks and *aguas frescas* (fresh fruit drinks) to tamales and *elotes* (roasted corn on the cob with all of the fixings).

I attended El Día del Trabajador Agrícola and many other events during my fieldwork. All had one thing in common: rows upon rows of pop-up shade tents sheltering health and social service organizations and outreach workers from nonprofits and city, county, and state agencies. Everyone had a folding table strewn with informational brochures, educational activities, and freebies, like plastic safety whistles, pencils, key chains, and magnets emblazoned with help hotline numbers. Many (though not all) of the outreach workers were bilingual and eager to share information about housing and food assistance, im/migrants' rights, migrant education, and even entrepreneurship programs for farmworkers who wanted to leave the fields to start their own small businesses. The California Department of Pesticide Regulation (CA DPR) distributed bilingual comic books illustrating pesticide drift scenarios and steps farmworkers could take to reduce their exposure, and they handed out wallet-size cards with an 800 number to report pesticide drift incidents and illnesses. Chapter 3 described some of limits of pesticide policies and interventions, many of which place the burdens of responsibility on farmworkers.

Upon entrance to the event, folks received a sheet of paper that looked like a passport. At each table an outreach worker would stamp the passport after visitors engaged in a conversation, asked a question, or picked up a brochure. Throughout my fieldwork I collected two file boxes full of such materials on topics ranging from domestic violence to what to do if the police or immigration officials showed up at one's door.

After filling their "passports" with stamps, folks could then enter *la rifa* (the raffle), a very popular pastime. Prizes included bicycles and desktop computers, bulk supplies of toilet paper and laundry detergent, and gift baskets filled with food items like canned meat and boxed macaroni and cheese. The nonperishable grocery items lingered in the raffle area as the opportunities to win dwindled. Food insecurity is high in farmworker households, but many did their best to lovingly prepare and savor home-cooked and culturally comforting and familiar meals.

The use of these "passports" to drive outreach efforts felt cruelly ironic, although the organizers probably meant well by it. A true passport, with its accompanying visas and stamps, is an official government document; it systematically and selectively grants and denies individuals the right to travel and access work opportunities depending on their country of origin and its political relationships with other nations. These are rights that many undocumented im/migrant farmworkers are denied. Even with a Mexican passport, which could sometimes be secured through a long and expensive process via the closest Mexican consulate, in San Jose, California, travel proved costly and difficult. Undocumented farmworkers couldn't just leave to visit their pueblos and *ranchos* (villages) in México. They did not get to visit their family members for holidays, to help with the corn harvests and other community events, or to say farewell to loved ones as they lay in their deathbeds. The im/migrant farmworkers I met had gone to great lengths to come to the United States—not for a better life, but to work to be able to feed their families.

These faux passports, laden with hidden meanings and unseen power, were used to move people around, from booth to booth, service to service. This paralleled what I witnessed while accompanying farmworkers to various health, legal, and social service appointments or assisting them with phone calls and paperwork. When one agency or organization could not help, people were often sent along to yet another, with no guarantees that their needs would be met.

In the middle of the park, La Clínica Campesina occupied a large central area shaded by several red pop-up tents. As a Federally Qualified Healthcare Center with branches throughout the region, it provided health care and other services to many farmworker families and other uninsured or underinsured individuals. I joined a growing line of a dozen farmworkers to undergo free blood sugar and blood pressure screenings. Nurses and medical assistants swabbed our fingers with alcohol and after a deft needle prick, squeezed drops of our blood onto thin paper strips. In mere seconds, the portable blood glucose monitors registered my "normal" results, which were scribbled onto a bilingual pamphlet featuring diabetes prevention measures and advice: *tome más agua, coma frutas y verduras, limítese a dos tortillas por día, haga ejercicio físico* (drink more water, eat fruits and vegetables, limit yourself to two tortillas a day, exercise).

I watched as the men received their results. Some smiled with relief. Others' faces grew stiff with shock and concern as medical assistants explained the potential consequences of unmanaged and untreated diabetes. As one nurse on duty told me, "Many people screened at these events have no idea they're diabetic. They get tested and their blood sugar is dangerously high. We tell them to go to the hospital immediately! But not everyone does." For some, these community screenings are the only interaction they have with a health care provider all year.

An announcement interrupted the music. The crowd clapped half-heartedly after one of the emcees spoke with embellished emotion over romantic Mexican guitar music. She professed the sheer economic worth of *los campesinos/as*: "It is they who make everything in agriculture possible!"

Some young men sitting near me on a patch of grass listened and smirked. I wanted to ask them what they thought. Did they feel appreciated? I held back, feeling unsure of myself and sensing their uneasiness about me, a white woman with a pen and notebook in hand; I was a strange and suspicious sight in a place like Greenfield.

I stashed my notebook away and headed back to my car. In the parking lot some Greenfield policemen interrogated an Indigenous farmworker and his teenage son. The cops stood in intimidating postures, chins up, chests out, hands resting on their guns and nightstick- holstered hips. A cooler lay on the ground with a few open cans of beer. One officer asked the dad, in English, to dump the beer. Another cop forcefully kicked a can over, its contents spilling onto the parched ground. Visibly anxious, the dad paced, eyes darting left to right, his hands frantically rubbing his face. His son stood off to the side, arms crossed, body slouched, head hanging low, eyes downcast.

Nearby a fellow police officer, out of uniform, and his wife chatted casually with uniformed officers about what a family-friendly, safe and fun day it had been. I thought to myself, For whom?

HOW DO WE CARE? AND WHY?

I started to put the mixed scenes, services, and emotions of the Día del Trabajador Agrícola event into a broader context. Hundreds of nonprofit organizations, private foundations and philanthropies, and local and state agencies have programs that aim (or claim) to support thousands of farmworkers and their families in the Pájaro and Salinas Valleys and throughout California. Many of these efforts are underfunded, understaffed, overwhelmed, and face heightened pressures to serve as many (who qualify) as possible in order to keep getting funding and thereby keep their doors open. They do this seemingly never-ending work in a climate characterized by anti-im/migrant sentiment and increasing state and federal fiscal austerity toward social welfare and health care that serve the impoverished and vulnerable. The multitudinous challenges that affect the lives of their farmworker clients and patients are not so easily reconciled during an office visit or with the receipt of an emergency food supply or an informational pamphlet.

Over the course of my fieldwork, I engaged with many nonprofit and social and health services professionals who, in earnest, did what they could to help farmworkers. Yet, more often than not, these care workers were consumed with paperwork and reporting requirements rather than engaging and supporting communities in direct service and social change work (Kohl-Arenas 2016; Shaw 2009). The passage of the Affordable Care Act sought to close some of these gaps, but many im/migrants remain excluded due to their income and/or citizenship status (Mulligan and Castañeda 2018; Horton et al. 2014).

Some philanthropic and nonprofit organizations in the Pájaro and Salinas Valleys are funded by the very agribusinesses that contract farmworkers to work the fields for poverty wages. People who had been involved in farmworker advocacy efforts felt that new industry interest in employee wellness was "better than nothing." They saw the availability of agriculture-funded farmworker health programs and resources as a vast improvement from past situations and far better than what farmworkers would (or would not) receive in México. Some agribusiness firms ran employee wellness programs with nutrition education and exercise classes. Others opened private clinics exclusively for employees and sometimes their families. They appointed higher-level corporate staff to sit on the boards of community nonprofit organizations. They awarded scholarships to the college-bound children of farmworkers or endowed libraries and buildings at public universities. They sponsored community events and plastered their logos on big banners hung above the streets. They donated massive quantities of fresh produce to food banks and schools. They were everywhere.

These acts do not come solely from a place of care and kindness. The history of philanthropy in the United States suggests that in many cases, such endeavors are tied to corporate profit motives. In keeping low-wage workers calm, contented, and modestly satiated, and by performing highly conspicuous acts of generosity, businesses attempt to foster rapport and trust in the communities where they operate and can often avoid costly worker strikes and slowdowns (Ahn 2017; Wark and Raventós 2018). Routinely I heard that agribusiness workplace wellness programs were good for workers and good for business. Companies could save money on training and insurance expenses and receive tax breaks for their contributions, donations, and volunteer hours. Positive publicity, and community, customer, and worker loyalty and pride, paid off in innumerable other ways.

The anthropologist Dinah Rajak has observed how the corporate social responsibility programs of mining companies in South Africa take interest in the health and welfare of their workers, at least as far as workers' HIV statuses affect productivity and the corporate image. As Rajak notes, "the promise of a confluence of efficient business and caring corporation, which combines moral imperatives with, as [French sociologist] Mauss put it, 'the cold reasoning of the business, banker, or capitalist.' Relations between employer and employee are being transformed as a result of corporate healthcare programmes, creating connections between the personal realm of sexual conduct and family life and the political economy of global capitalism" (2010, 552; citing Mauss 1967, 23). The existence and prevalence of nonprofit institutions, philanthropy, and charity corresponds directly to the retraction of state support for basic needs in society and the rise in power, control, and authority of corporations (Ahn 2017; Kohl-Arenas 2016; Wark and Raventós 2018). These arrangements can influence how we think about what it means to be civically or anthropologically engaged. Corporate and community programs often take on the feel and form of nonprofits, shaping how and why people care, what we care about, and what we think we can or cannot do to support the health and well-being of farmworkers and others. Because of how they are structured, funded, and ideologically framed, there is only so much that formal institutions and corporate programs of care can accomplish. The "cheapening" of farmworkers' lives, livelihoods, health, and well-being, as well as the food the farmworkers produce, is a fundamental aspect of how agribusinesses profit (Patel and Moore 2017). More significant rearrangements and shifts in power would reduce corporate profits but perhaps do more to improve worker health.

Sometimes in contradiction to the consensus of "better than nothing," I found myself thinking, Is any of this really good enough? My observations of well-meaning, underfunded, corporate-sponsored, politically constrained, nonprofit, and other institutionalized forms of care led me to think more deeply about how caring is defined and enacted in diverse, complicated, and sometimes contradictory ways in im/migrant farmworker communities (Dilger, Huschke, and Mattes 2015, 1–2; Scheper-Hughes 1992). Upholding the Hippocratic oath to "do no harm" in im/migrant communities under systems of market-based medicine and agriculture is fraught with contradictions and difficulties (Rylko-Bauer and Farmer 2002). Some medical doctors in the state of Georgia, for example, have resisted laws that prohibit spending state funds to help or care for undocumented people. They acknowledge and respond to how inhibiting access to care and other punitive policies exacerbate im/migrants' multilayered health issues (Kline 2019).

Medical anthropologist Susan Shaw documents how the goals and values of community health activists and advocates are also selectively consumed by the state, turning grassroots and social movement demands for respect, recognition, dignity, and justice in health care and social services into a series of obtuse deliverables. These can include generalizable and measurable objectives of what constitutes "health," such as a collection of blood glucose or blood pressure levels, or the development of health interventions that focus only at the level of individual behavioral changes, like eating more fruits and vegetables or getting more exercise, while addressing very little else (Shaw 2009). The initial recognition of problems followed by formal institutionalization of communities and their demands for health justice may result in some desired and tangible outcomes. In the Pájaro and Salinas Valleys, these have included the construction of new neighborhood clinics, the hiring of bilingual health care providers and educators, the installation of pedestrian and bike safety infrastructure, the banning or heightened regulation of some pesticides, and the provision of personal protective equipment at some (but not all) worksites. These small victories, while important, can also be tenuous, incomplete, imperfect, or impermanent. They may also inadvertently produce gaps or shortfalls resulting from the fact that, after a win, attention and urgency shifts to the next crisis: a system of acute care instead of chronic and continuous care.

Urban sociologist Michael Lipsky notes that the logics of social service and public institutions, whether run by the state or privately, emphasize the management of problems through quick fixes and interventions instead of resolving them permanently or working proactively to prevent them. This is especially true when proposed changes and transformations threaten the comfort, wealth accumulation, and status of the powerful (Lipsky 2010) or, in this case, agribusiness as usual. The state and agribusinesses also routinely mobilize to tone down or slow down continued calls for community health and environmental and social justice (Shaw 2009, 186–187), maintaining that resources are limited or that people's concerns aren't valid or worthy of serious consideration.

Anthropologists, too, can be caught in these binds. Our formal ethical codes do not prepare us well for navigating and negotiating different moral expectations and ethically fraught situations or for coming face-to-face with intense forms of social and environmental suffering for which there are no immediate, if any, forms of redress (Bourgois 2006, 2009, 2010; Chin 2013; Dilger, Huschke, and Mattes 2015, 1–2; Liboiron, Tironi, and Calvillo 2018; Maeckelbergh 2016, 211).

During my fieldwork I often felt frustrated by the structural, spatial, temporal, and emotional limits of doing engaged ethnographic work in a place with so many contradictory and complicated values and norms guiding so-called caring companies, employers, clinics, policy makers, agencies, and authorities. When the police, in the name of public safety, harshly enforce open container laws, or similar low-level offenses, it creates distress, trauma, and financial and social hardship for a farmworker family with an arrest or fine. Nonprofit clinic workers screen farmworkers for diabetes and give them nutritional and lifestyle advice but rarely engage or organize communities around strategies to end food insecurity, poverty, air and water pollution, and the workplace conditions connected to farmworker health disparities. When representatives from the CA DPR send outreach staff to distribute bilingual brochures about how farmworkers can protect themselves from pesticides, they discursively and practically neglect the multilayered toxic realities farmworkers face at work and at home. When a judge and Social Security auditor interrogated Armanda, whose story is recounted below, they followed the rule of law and at the same time negated her legitimacy as a person who had been wronged and harmed at work. There are also laws that explicitly prevent nonprofits and some state actors from engaging in systems-changing work. Those who want to participate have to be careful to do so on their own time and not as representatives of the organizations or governments they work for (Kohl-Arenas 2016).

Some might respond that these staff, health care providers, public servants, corporate philanthropists, and anthropologists, in their varying roles, are just doing their jobs and that we cannot expect them to do more. Perhaps physician-anthropologist Paul Farmer's idea of pragmatic solidarity (2003, 2004), of doing what's possible and realistic, and Nancy Scheper-Hughes's "good enough" anthropology, of doing the best we can in difficult circumstances with insufficient resources (1992, 28), are ways of coming to terms with our own limits as ethnographers, concerned citizens, and ordinary humans (Shaw 2009; Williams 2015). I routinely feel this way in my own work.

Pragmatic thinking and acting may be conditioned by desires to respond quickly to a known problem in order to mitigate harm or to *do something* (Farmer 2004). Institutional cultures; our social, geographical; and temporal locations and positionalities; our professional codes of conduct and ethics; and our personal values shape our perceptions of what is real, what is possible or doable, and what is right and just (Bourgois 2009, 2010; Chin 2013). This creates tensions for people who care about or for communities. Do we focus on providing urgently needed social services, being supportive more generally, or acting for sustained social change in precarious and trying sociopolitical and ecological circumstances (Kivel 2017; Kohl-Arenas 2016)? Our desires as engaged anthropologists to contribute to the health and social and environmental justice goals of communities in meaningful ways also often come up against a sense of hopelessness, despair, and exhaustion. The challenges communities face sometimes feel irresolvable or impossible (Duncan 2018; Heidbrink 2018; Horton 2016b; Liboiron, Tironi and Calvillo 2018; Shaw 2009; Stuesse 2016). But limiting our work to what's practical or realistic just doesn't seem "good enough" (Durham 2016).

ANTHROPOLOGICAL ACCOMPANIMENT AS CARE WORK

Medical anthropologist Whitney Duncan describes engagement with im/migrant communities in a post–Donald Trump world as a form of accompaniment. Im/migrants themselves routinely evoke the Spanish word *acompañamiento* to describe the supportive roles of outsiders as well as the care and mutual aid im/migrants themselves perform for one another as together they deal with the harms of deportation and family separations. Activist historian and attorney Staughton Lynd thinks about accompaniment as a relationship between "two experts . . . exploring the way forward together," or people with complementary skills, insights, and experiences that can be comobilized in social justice work (2013, 4). The practical and emotional meanings of *accompaniment* are even clearer in the Spanish *compañero/a* or Portuguese *companheiro/a*, implying partnership, fellowship, solidarity, and friendship combined (Pine 2013; Scheper-Hughes 1992, 1995).

The social and emotional relationships of accompaniment go beyond what's institutionally conscribed or allowed. They push us to create alternatives that rupture standard and harmful social and research arrangements. Pragmatism aside, Farmer (2011) has also invoked accompaniment to describe the critical reimagination of health care practice as something that should go above and beyond the impersonal interactions and short-term involvements characteristic of clinical encounters, public health interventions, or humanitarian aid missions. Accompaniment in health care and social change work requires long-term commitments, sustained relationships grounded in "physical proximity" and "mutuality" and privileging the needs of the most vulnerable members of our society (Farmer 2011, citing Goizueta 2009 and G. Gutiérrez 1973).[1]

As exciting, transformative, or idealistic as it sounds (and is), accompaniment is not always easy, nor is it inherently rewarding. It is time-consuming and can deplete one's energy and spirit. As Duncan observes, "As a model of resistance, *acompañamiento* highlights . . . suffering and solidarity. . . . At the same time, it highlights the limits of solidarity and support in the absence of structural change"

(2018). Even with its shortcomings and heartbreak, accompaniment is necessary humanizing work. I came to see its potential as a model of activist research and relationship building that contrasted institutionally or corporately constrained forms of caring.

In this chapter I share some of my experiences accompanying farmworkers. I have come to think of accompaniment as a more politicized way of applying the classic ethnographic method of participant observation. Accompaniment allowed me to engage more deeply by following the leads, threads, and concerns expressed by im/migrant farmworkers rather than merely following my original research protocol. I want to argue that even with its emotional heaviness and practical limits, accompaniment holds a lot of imaginative power when it comes to communities' and activists' efforts to seek, create, and build alternatives to current exploitative, extractive, and toxic ecosocial relationships throughout the food system (Khasnabish 2016). Pragmatism, hopelessness, despair, "good enough," and "better than nothing" are human feelings and experiences, but they do not always generate the energies that become driving forces for change. Building on this I will, in chapter 5, describe people in familial, community, and institutional relationships with farmworkers who engaged in solidarity even amid significant and toxic challenges, political barriers, and social, occupational, class, generational, and educational differences. In different ways they taught me to model my own research around the cultures of care and concern found within farmworker families and community advocacy and activist circles. It is these every day and largely invisible acts of care, these emotional and social labors, that I believe represent significant yet underappreciated forms of activism.

ARMANDA

Armanda felt a sharp pain in her back while washing freshly harvested vegetables. The job involved heaving thirty- to sixty-pound plastic crates filled awkwardly and unevenly with produce, dunking them into a tub of cold, bleach-laced water, lifting the now forty- to seventy-pound wet boxes, and placing them on racks to dry—over and over and over again, for up to ten hours a day, five or six days a week.

The farm owner, Chuck, a white man, stationed specific workers in the same repetitive jobs day after day. He replicated the harvest management strategies and ideologies of larger commercial operations in the Pájaro and Salinas Valleys, which divided workers into groups and tasks by gender and ethnicity (see Cartwright 2011, Holmes 2013, and Horton 2016b). He lorded over the workers, pushing them to work as hard and as fast as possible. This made them especially susceptible to injuries.[2]

Armanda, her partner, Ismael, and some of their compañeros/as rented a house together. Whenever I visited, we sat around the kitchen table and they exchanged injury narratives and stories of workplace stress. The back injury that Armanda sustained while working on Chuck's farm was not limited to her physical body. It exacerbated a long-standing deep depression, intersecting with emotional traumas she endured after being gang-raped as a teen and experiencing domestic violence from a previous partner.[3] Not working heightened Armanda's day-to-day anxieties and pain. Her mind raced, but her body felt numb and heavy.

How would she support her children and family members in both México and the United States? Armanda's inability to work put additional pressure on Ismael, who had to work harder and for longer hours on the farm. He also injured his back while carrying heavy metal irrigation pipes but worked through the pain. Whenever they asked for permission to take a leave or get off from work early to see a doctor, Chuck reprimanded them and accused them of being lazy.

Initially Armanda and Ismael resorted to home remedies and treatments. These eased but did not stop the pain. As Armanda explained, they did not want to sacrifice their limited funds on a doctor's visit:

A veces [Ismael] no se puede agachar. Y a él le digo que vaya al doctor, que paga, [y me dijo,] "No, pero no hay dinero." Yo tengo que aveces en la noche voy subiendolo, poniendo agua frio. . . . Hay veces en que creo que sería buena que se lleva a un doctor para que haga un estudio para que miran si . . . haya lastimado o fractuado un hueso o algo. Pero él, pues dice, "No puedo pagar, porque no hay dinero." Y por eso no le dijo al patrón que seguía enfermo, porque él se va a enojar. A lo mejor que se queda asi.

Sometimes [Ismael] can't bend over. And I tell him to go to the doctor, to pay for it, [and he tells me,] "No, but there's no money." Sometimes at night I have to lift him up and put cold water on his back. . . . Sometimes I think it would be good to take him to the doctor, to do a study to see if he's been injured or if he's fractured a bone or something. But he, well, he says, "I can't pay, because there's no money." And this is why he didn't say anything to the boss that he's still sick, because he'll get angry. Maybe [Ismael] will stay this way [injured].

When Armanda first notified Chuck about her back injury, he told her to take an ibuprofen. He cursed at her in rough, broken Spanish. She kept working, fearing that if she complained or sought legal help she would lose her job. Carlos, as the workers somewhat jokingly referred to their white boss, scoffed at Armanda: "¡Muchos problemas contigo¡" (You have so many problems!) When her pain did not subside, Carlos was hostile, instructing Armanda to go to the local nonprofit clinic on her own time. There a doctor performed minimal evaluations (X-rays, asking questions about pain), prescribed muscle relaxants, and signed a note granting Armanda two weeks of rest, after which she could return to work. The same pattern followed for Ismael.

Eventually they each filed workers' compensation claims, but this only led to retaliation and harassment. At this point Armanda's depression intensified. Her only motivation to get out of bed each day was to attend to her young daughter Nena, a temperamental toddler with an array of special needs including asthma and learning difficulties.

At a different clinic, doctors recommended that Armanda take muscle relaxants and painkillers. These made her dizzy, fatigued, and nauseous. To address her complaints of depression and anxiety, the staff referred her to the psychiatrist, who urged Armanda to go on antidepressants and to collect Social Security Disability Insurance while she took some time off to deal with her emotional problems. Armanda complied with their recommendations, but this led to yet more problems and intensified vulnerabilities.

Armanda's borrowed Social Security number, used to process her disability and health care paperwork, came up in an audit. I drove Armanda to her court date and sat with her during the hearing. The judge and the auditor, who testified over speakerphone, demonstrated frustration and skepticism about Armanda's truthfulness.

We took a five-minute break to consult with the translator on duty, to ensure that Armanda fully understood what was happening. Through tears, she pleaded to the judge that she was being honest about her situation. That she really was disabled, and that she feared what would happen to her children if anything happened to her. The judge's facial expression grew more sympathetic, but also exasperated. Armanda's honesty was not the only thing to go on trial in that courtroom. Her actual existence and legitimacy as an injured and legally and medically wronged person also came under close scrutiny.

In light of Nena's precarious health and the fate of Armanda's other children, ages seven to fourteen, who lived in México at the time, the judge ruled that she would not have to pay the state back the money she had collected while on disability, totaling nine thousand dollars. More critically, she would not (immediately) be reported to Immigration and Customs Enforcement for arrest. The judge warned her, however, that admitting her undocumented status on the record could lead to repercussions later on, especially if she found herself in a courtroom again.

Ultimately, after years of back-and-forth between doctors, lawyers, insurance companies, and the farm, neither Armanda nor Ismael found complete relief for their ailments or justice at their jobs. They've tried starting their own business, but this involves navigating yet more bureaucracies, paperwork, and rules that are often unfriendly and unaccommodating to Spanish speakers.[4] Accompanying Armanda and her family throughout their medical and legal ordeals gave me perspectives about gaps in care and allowed me to witness the systemic shortcomings and barriers to care that perpetuate harm in farmworker communities. Accompaniment is not an ultimate solution to Armanda and her family's multilayered problems. It did, however, politicize the ethnographic act of participant observation and turned everyday acts of caring and observing into forms of activism that contrast with the social distance and indifference Armanda and others encountered in their daily lives.

WAITING IN LINES

It is not always easy for people who need food to access it, even when there are food banks and other organizations locally that frame themselves as helping, feeding, and caring (Carney 2015). I accompanied two Zapotec-speaking sisters, Hortensia and Oralia, both pregnant at the time, and their children to the food bank in Salinas. They each lived in small single-family homes with two to three bedrooms a piece in rural northern Monterey County, along with three to four (and sometimes many more) additional farmworker families all crammed together. Landlords also parked decrepit and decaying RVs in the backyards to shelter additional people. Upwards of twenty-five people could be sharing a single bathroom. Despite charging astronomical rents, landlords routinely neglected to provide farmworker households with working stoves and refrigerators. Technically, these items are required under California tenants' rights laws. When things broke down—and they often did—it fell on farmworker renters to find a quick fix, replacement, or go without.

It had become increasingly hard for the sisters, both of whom endured difficult pregnancies, to work in the fields. Before becoming pregnant with her third child, Oralia developed back pain from the repetitive stooped labor of harvesting strawberries. A debilitating depression and anxiety followed that intensified during and after the births of her two youngest daughters. She fretted about how her family would get by if she could no longer contribute financially. Oralia's husband, Santiago, also felt a lot of pressure. The whites of his eyes routinely turned red, splotched by tiny burst vessels as his blood pressure soared. He wouldn't go to a clinic; their oldest daughters, at ten and twelve years, had their own health problems and health care for the kids took priority. Sometimes Oralia and Santiago fought. He would leave the house for a few days to clear his head, causing Oralia to become even more panicked.

I took some time to investigate where these families might be able to get some food assistance. I hoped that such assistance might alleviate some of their anxieties, but figuring this out with them proved to be a lot of work in and of itself. I had seen the distribution lines in town in Watsonville, which would have been easier for the sisters to get to, but since they were not residents of Santa Cruz County, they were ineligible (or so we were told). Monterey County had distribution sites, but they were only open while most folks were at work. Both families had cars, but their husbands worked long hours in the fields and needed them.

After making several telephone calls, I found that the next distribution would take place in Salinas on a Thursday, twenty miles away from where the families lived. It took about an hour to load everyone, kids and all, into Hortensia's van and then make the drive south. They asked me if I could do the driving; the rapid speeds and erratic actions of drivers on Highway 101 made the sisters nervous. If they had taken the bus, the quest for food would have taken the whole day, in addition to bus fare for the sisters and all children over five years of age.

At the distribution site, which was right at the food bank's warehouse, an intake volunteer greeted everyone in Spanish. Before they could get in line to receive any food, they had to register, sharing information about their household sizes, monthly incomes, phone numbers, and street addresses. Neither Hortensia nor Oralia could read or write.

Oralia became tense. Would registering her family at the food bank put them at risk of being reported or deported? Likely, the information would be used by the food bank internally to document the demographics of recipients and the quantities of food given away. This data could be used to apply for future funding and support.

At that time, and into the present in California, there is no requirement for people to provide proof of citizenship at nonprofit food distribution sites, nor is it lawful for food bank and pantry personnel to ask. Yet at other food distributions in California, volunteers and employees have been known to police recipients' access to food by requiring intensive intake questionnaires, surveilling people during their visits, and policing the amounts and kinds of food they take (Fisher 2018).

The food bank intake volunteer reassured Hortensia and Oralia that the organization would not share their information. The sisters consented, and I helped them fill out the forms. After making it over this hurdle, we waited in line for about a half hour along with hundreds of other families. Some had walked over from nearby neighborhoods in Salinas, armed with their own rolling grocery carts and reusable bags to carry home their rations.

Another volunteer handed each sister a carton of eggs. Oralia wondered if she could have another one since her family was so large, but she was denied. Only one carton per family, the volunteer said.

We proceeded to the next area, picking up five-pound bags of rice, pinto beans, and a box of assorted canned and dry goods. At the next station were several large cardboard bins filled to the brim with assorted loaves of white bread, surpluses donated by area grocery stores. People could take as much of this as they thought they could eat. The last line featured large bags of potatoes and apples. The woman monitoring this section let Hortensia and Oralia have an extra bag of apples each for their kids. I was also offered a bag of apples, which I took and gave to the sisters.

We headed to the van, loading up the modest provisions and the children. Hortensia and Oralia both lamented that these items, while appreciated, would barely get them through the next week. The Salinas distribution only happened once a month. Folks could visit other sites in other towns and cities hosted in other weeks. It took a lot of cost, time, and energy—from the fuel in the car, to the time investigating where to go, to then driving there, to the emotional intensity of weighing whether or not registering their families would cause them harm later on—to receive such a small amount of food.

When visiting farmworker households, I took note of what people cooked on stovetops, what sat in kitchen pantries and storage, and the prepared meals or fresh fruits offered to me. I routinely observed men, women, and children carrying the telltale recycled mesh or plastic produce bags filled with staples like potatoes, lettuce, cabbage, onions, carrots, apples, and oranges, as well as beans and rice and canned goods. For farmworker families, food bank foodstuffs helped fill voids caused by low household income, especially in the lean winter months when there were no harvesting jobs. Like clockwork, Central Coast food bank employees and volunteers prepared for increased traffic at their distribution sites from November through March, while the strawberries "slept" (*están durmiendo*).

THE FOOD POLITICS OF CARE

I do not share this story and critique to dismiss the many staff at food assistance and other nonprofit and community-based organizations who are deeply concerned about the health and welfare of their communities. Indeed, some may very well share in the critique of inadequacies (Fisher 2018; Kohl-Arenas 2016; Shaw 2009). Food pantries and banks do help farmworkers and countless others, but food assistance programs in general are under attack at the time of this writing. Under the Trump administration's new mandate, im/migrants' prospects of citizenship are being denied if they or any of their citizen family members have ever received any sort of federal assistance for food, housing, or other things.

As author Andy Fisher explains, on the surface, food assistance programs frame their work as helping people with limited income or resources get by, but these institutions also represent yet another way that agribusinesses' profits are subsidized. The history of food banks dates back to twentieth-century philanthropic endeavors, which sought to suppress workers' challenges to abhorrent and deadly labor and living conditions. Strategic gifts of food, and school supplies, and the opening of company clinics, made life a little more bearable, fostered a sense (however superficial or temporary) of good faith between workers and the elites who owned and managed the companies, and cultivated trust and respect for companies from the general public (Ahn 2017; Besky 2018; Fisher 2018; Patel and Moore 2017; Wark and Ravéntos 2018).

Questions about why separate tiers of care and support exist for impoverished people in the first place don't get posed upon receiving a conspicuous gift from someone with more power. As Scott Anger and Sasha Khoka note, "food banks . . . still primarily hand out packaged food," as Hortensia and Oralia experienced, "but they're working to increase their donations of fresh fruits and vegetables so residents can share in some of the bounty they helped to harvest" (2013). The eggs, apples, and potatoes Hortensia and Oralia picked up aren't just the result of a generous donation by a food producer or retailer. They also result in a tax break, the avoidance of dumping fees, and good publicity.

In Anger and Kokha's documentary series *Hunger in the Valley of Plenty* (2013), a volunteer from the Central California Food Bank, which serves parts of the Central Valley, reflects upon the young children accompanying their parents at the distribution lines who light up at the sight of fresh strawberries. These are sometimes donated by berry grower-shipper companies from surplus or slightly imperfect harvests. As the volunteer explains, "When I see those families come by the line and come pick up the food, I just look at the children, their faces. When they see those strawberries, to them, it's like a candy; it's like a treasure for them" (Anger and Kokha 2013).

The optimistic symbolism of the strawberry came up again at a 2016 fundraising gala for the Central California Food Bank that I attended with some colleagues less than twenty-four hours after it had been announced that Donald Trump had won the U.S. presidential election. People convened at a stately banquet hall in downtown Fresno with marble floors and tall white pillars that had once been a grand hotel ballroom. Members of a local ceramics club greeted us at the door, and each guest received a handmade "empty bowl" representing one of the hundreds of thousands of people in the Central Valley who face food insecurity and chronic hunger.[5] A table full of silent auction items included an assortment of gift baskets, bottles of wine, season tickets to minor league baseball and soccer games, and an opportunity to practice "tactical maneuvers" and "testing your skill on the [shooting] range" with the Fresno Police Department's SWAT team (valued at a thousand dollars).

My tablemates and I felt a range of emotions that evening, but mostly rage. The woman who handed me my empty bowl and the folks standing around the cash bar and eyeing silent auction items seemed carefree. Upon taking the microphone to welcome everyone, the CEO of the food bank encouraged attendees to put politics aside for the evening and focus on the goal of giving so that others would not go hungry. For him, food banks and strawberries, even amid times of conflict, represented hope: "When it comes down to it, at the end of the day, we provide hope. When you see the people in line and they leave with a spring in their step, and when you see children in the lines leave with a little box of strawberries, it feels like Christmas. . . . At the end of the day, we provide hope."

I would not describe the emotions of Hortensia or Oralia, as they waited in line with their children for food and as they faced many other health and life challenges, as hopeful. While the food they received was helpful and needed, the sisters' overall circumstances—underemployed, injured, stressed out, sick, and living in poverty—changed very little with or without food assistance. The fact that they had to immigrate to the United States to feed their families on both sides of the Mexican-U.S. border is an intensely political issue. So too is the sisters' reliance on places like food banks and their overall economic and social subjugation as undocumented farmworkers.

Decadent fundraisers like the one I attended in 2016 regularly highlighted the contributions of area agribusinesses. Luncheons honored record-breaking or long-term donors and volunteers, including many with agribusiness affiliations. Local newspapers routinely covered food bank distribution statistics, ranging in the millions of tons, with substantial increases year after year. Maybe an especially generous patron would win the chance to practice shooting with the SWAT team, with state weapons that in all likelihood would be aimed at yet more impoverished and disenfranchised people.

Many who do the work of hunger relief—high-profile donors, board members, or food bank directors and coordinators—carefully reassure the public that their work and the gifts they receive are benefiting people who need it and deserve it. One food assistance organization, SHARE, serving the Pájaro and Salinas Valleys often emphasized that food distributions were not just for homeless people; working and middle class people also depended on food banks and pantries to get by. Farmworkers, who harvest the produce that gets distributed and who may also be the recipients of that same food in rural California communities, were rarely mentioned. Agribusinesses distance themselves from their roles in fostering the root inequalities that perpetuate disease, illness, poverty, and hunger among their own employees. This includes their participation in transnational processes that displace Mexican farmers and their families (including the many whose stories are told in this book) from their lands and reproduce poverty in both México and the United States (Zavella 2001, 2011, 2016).

Meanwhile, the inevitability of poverty as something that "always has been and always will be" is reiterated during public events celebrating conspicuous giving. Sociologist Erica Kohl-Arenas (2016) demonstrates how in California the work of foundations, philanthropies, and nonprofits—with support from their corporate funders, including agribusinesses—perpetuate myths about how social progress is achieved through charity-based poverty eradication and community empowerment efforts in the United States and elsewhere. Programs emphasizing civic participation, education, and capacity development for poor people are most likely to get funded by foundations. These activities put a lot of the responsibility for change, growth, and health on communities themselves and mask how the political power and concentrated wealth of agribusiness corporations, including those involved in charitable or philanthropic endeavors, continue to produce the very conditions of socioeconomic inequality and political marginalization that keep poor people, including im/migrant farmworkers, poor. While mitigating some suffering and achieving some desired development outcomes for communities through the establishment of new parks or housing, creating health care plans and wellness programs, opening clinics, or helping select students get through their first year in college (see chapter 5), such projects and programs routinely fail to address the roles of corporate donors in perpetuating and exacerbating conditions of ill health, toxicity, and social inequalities. This effectively depoliticizes farmworker health and wellness.

I see practices of accompaniment as a challenge to this depoliticization. Accompaniment, or being a *compañero/a*, enables us to witness and experience things differently from the emboldened and top-down position of a benevolent donor, employer, or a stretched-thin service provider. It models how care can be an explicitly political act.

ACCOMPANYING ANICETA

In the fall of 2011, residents of Greenfield—including a former mayor, community organizers, youth, and farmworkers—traveled north to the touristy beach city of Santa Cruz to participate in a panel discussion, "The Truth about the South Salinas Valley." They shared their stories and experiences with a mostly sympathetic, white, and politically progressive audience. Aniceta, a thirty-five-year-old Triqui woman, and her two youngest children, Pablo and Vicente, then ages eight and eleven, sat in the audience.

They had hitched a ride with Lola, who sat up front among the other panelists, who were all second-, third-, and fourth-generation politically active Mexican Americans with nonfarm jobs. Lola and her peers described how they had left the fields to pursue education and higher-paying jobs after the passage of the Immigration Reform and Control Act in 1986. These opportunities led to their participation in community organizing and local politics. Some had also run for positions on city councils or school boards.

Aniceta's story stood in stark contrast to those of the panelists. She and four out of her five children had migrated to the Salinas Valley in 2000, among thousands of others from west central México's Indigenous rural peasant communities fleeing rural economic instability, chronic joblessness, cartel and state violence, and interethnic conflict (see A. A. López 2007; Rivera-Salgado 2014; Stephen 2007). With few viable pathways to legal residency or citizenship, farmwork remained the only option for newer im/migrants like Aniceta, her husband, and their older children.

One day, while harvesting peas in fields buttressing the Salinas Valley's sun-dried eastern Gavilan Range, Aniceta's vision blurred. She felt dizzy and weak.

Sweat drenched her face and shirt, and her heart raced and pounded. She did not want to tell her supervisor what was happening. Earlier he had scolded her when she asked him which rows to harvest first. Both explicit and tacit hostilities were a constant. Aniceta knew she was not one of the *preferidas*, or female favorites, who received routine praise and the prime picking spots to maximize their piece rates.

Aniceta's coworkers coaxed her to sit under a shade tree for a bit. One *paisano* (compatriot) offered some of his water. She fell in and out of consciousness for at least an hour before a supervisor noticed.

The supervisor first took Aniceta to a gas station, forcing her to chug Gatorade: "Me dijo, 'Toma, toma!' Entonces, tomé y tomé." (She told me, "Drink, drink!" So I drank and drank.) From there, she drove Aniceta to an urgent care center in the next town, over an hour away. Horrified at Aniceta's now critical condition, one of the nurses called for an ambulance.

On the way to the emergency room in Salinas, another forty minutes away, Aniceta's heart went into irregular rhythms, then stopped. Paramedics and doctors shocked her back to life several times. "Me taparon aquí [*making a motion with her hand mimicking the placement of an oxygen mask*], y ya no supé nada, y después me revivieron con la plancha de electricidad" (They covered my face, and then I didn't know what was happening, and then they brought me back to life with the electric iron [Aniceta's way of describing a defibrillator]").

Aniceta's heart sustained permanent damage at thirty-five years of age. Her doctors installed a pacemaker, which she would wear for the rest of her life. Regular visits to cardiologists and other specialists became part of her routine. Months later, Aniceta found out she had type II diabetes. She endured chronic pains at her incision sites and elsewhere, debilitating depression and anxiety, dizziness, vomiting, blurred vision, sleep apnea, and heart palpitations. She could barely walk a block without getting fatigued, winded, and sweaty. Initially, whenever her pacemaker gave her a *toque* (shock) stimulating her heart back into regular rhythm, she would panic and rush to the emergency room in Salinas. The sensation made her want to die.

Aniceta insisted that before her illness, she had never had heart problems, and hardly ever got sick. She lamented her inability to contribute to her already struggling family, including her children and grandchildren in the Salinas Valley and her parents, siblings, nieces, and nephews in their *rancho* in Oaxaca. Her three oldest children, all young parents themselves, were overwhelmed with their own challenges: drug and alcohol addiction, intimate partner violence, raising small children on a farmworker's salary, and living as im/migrants in a politically hostile clime. They now had to contribute more to support their mother. Eventually, as Pablo and Vicente got older, they also helped take care of their mother.

Aniceta's husband, Bernando, already a heavy drinker, felt intensified stress from being stretched thin at work in the fields. He started drinking even more and became increasingly belligerent and abusive toward his wife and children. In one drunken fit, he attacked Aniceta and her eldest son with a knife.

Eventually Aniceta left him. She and the boys moved from place to place, including two garages, a women's shelter, and a series of decrepit apartments. I visited them at each of these places. Sometimes, when she ran out of money, her phone would be cut off and we would lose touch for months at a time.

Between 2011 and 2013 none of the many doctors she saw in community clinics, hospitals, or specialists' offices attributed Aniceta's heart condition to heat exposure or farmwork on the grounds that they could not be certain of the exact cause. Without a doctor's recommendation, Aniceta could not receive care or financial support through workers' compensation. She was also ineligible for long-term disability coverage (see Horton 2016b).

In February 2011, I accompanied Aniceta to a cardiac clinic appointment, giving her a ride and providing moral and some translation support as she let me listen in and observe. This became particularly difficult when it came to Aniceta's requests to have her doctors sign letters attributing her heart condition to workplace conditions and negligence. Her cardiologist was based in Salinas, an hour away from Greenfield by car and at least two hours by bus each way. Dr. Kolani (a pseudonym), while sympathetic to Aniceta's struggles, would not sign anything. He could not be certain that her heart condition was work related, but at one visit, he shared his suspicions that many of his patients under fifty suffered from cardiac problems that, at least in part, were due to the effects of toxic pesticides. Still, he was not in a position to prove it.

Aniceta sat upon the examination table in a light blue hospital gown following a cardiac ultrasound. Her heart had weakened further since her last checkup and now worked at 40 percent of its normal capacity. These technical diagnoses and legal definitions of "disabled" were not easy for Aniceta to understand, let alone accept. Even though plenty of nurses and medical assistants on staff spoke Spanish, many things remained lost in translation, since Aniceta's first language was Triqui, and because technical medical Spanish is distinct from the lay Spanish farmworkers speak. Dr. Kolani grew frustrated, suggesting that they had explained her diagnosis and care instructions countless times, but Aniceta insisted they hadn't. In sum, over the years, Aniceta's test results and exams revealed cardiomyopathy (weakening of the heart) as well as an arrhythmia (irregular heartbeat), and a diagnosis of congestive heart failure—all before her fortieth birthday.

We rode back to Greenfield to stop at the pharmacy. Upon Aniceta's request, I attempted to read the different prescription labels to try and figure out which *pastilla* (pill) addressed what problem, which pills might be connected to certain unpleasant side effects, and when and how often to take them. She toted her collection of pills around in her purse everywhere she went. Several times she'd dump them all out in front of me so that I could interpret the labels for her and relabel them in Spanish. In moments of despair and frustration, she would also

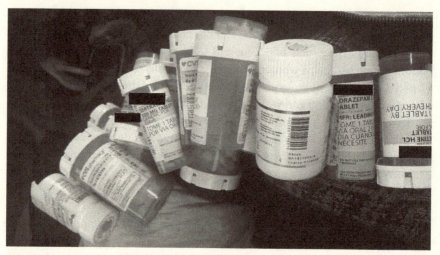

FIGURE 14. Aniceta's *pastillas* (pills). (Photo courtesy of Aniceta.)

text me photographs of her pills lined up in no particular order. Despite the region hosting a large Spanish- and Indigenous-language-speaking population, pharmacists did not always voluntarily provide Spanish-language labels or accessible explanations of what they prescribe.

For a while, a lawyer tried to pursue a tort negligence and personal injury case on Aniceta's behalf; however, this proved too challenging for her to keep up with, and with each move, notices from her attorney got lost in the mail. Eventually the California Occupational Safety and Health Administration fined the labor contractor who failed to call the ambulance on Aniceta's behalf the sum of five thousand dollars.

I accompanied Aniceta and some of her children to numerous appointments and through a nonstop whirlwind of life events. Sometimes, accompaniment looked and felt like a kind of de facto social work. I helped Aniceta read and interpret bewildering medical and legal forms that came in the mail. I wrote a letter on her behalf when county social workers from Child Protective Services investigated Aniceta after teachers expressed concerns about Vicente ditching school. They required her to attend parenting classes. Her sons struggled a lot with trauma, poverty, peer pressure, girlfriend problems, the lure of self-medication with marijuana and alcohol, and staying interested in and keeping up with their homework. Her daughters deal with abusive partners and the challenges of balancing motherhood with seasonal farmwork.

Vicente has since graduated from high school—the first in his family to do so. He, Pablo, and some of their cousins are considering college. On a tour of area colleges and universities that I arranged for them, one of Vicente's main questions was about how he would balance his studies with caring for his mom. Could he rent a Zipcar for weekend visits? In contrast, the tour guide explained that the rental cars were more useful for grocery runs or taking spontaneous road trips to San Francisco and other nearby attractions.

Throughout my research I was often, not so ironically, mistaken for a social worker—not by farmworkers (with whom I was very clear about my roles and objectives) but by passersby or health care workers who wondered out loud if I was a caseworker—*una trabajadora*.[6] In other instances, I ended up in these roles, sometimes voluntarily and sometimes involuntarily. Farmworkers often asked me for help navigating and negotiating their many health, legal, food, housing, diaper, and children's schooling needs, as well as their insecurities and uncertainties. I learned a lot about these everyday life course issues. After taking Aniceta or others to appointments, I would go home and look up laws and social services that might provide some answers or support—often coming up with little to nothing. In Aniceta's case, these questions were broad: Where could her sons get therapy? Where could her eldest daughter attend classes to earn her GED? Where could they find affordable and child-friendly housing?

It was often exhausting and draining work for everyone involved. The fact that so many folks asked me so many things told me a lot about the gaps—social, political, and practical—in formal systems of care and support.

Eventually Aniceta, too, learned the ins and outs of these systems as a consequence of her health conditions and her fraught circumstances as a disabled, Indigenous, unemployed, undocumented single mother of teenage boys. At one point she explored getting a business license to sell *artesania*—things she weaves and makes at home, when her back doesn't hurt, when she's not too sad, and when she has the energy. She remembers the skills her mother and grandmother taught her back in Oaxaca. This is the one job she could see herself doing, given her state of chronically declining health. A family business is something she could leave behind as a legacy to her sons. Aniceta harbored less of an American dream and more of an Indigenous Oaxacan binational dream.

Unfortunately, the friendship bracelets and handwoven *servilletas* (napkins) she made were hard to sell. Every once in a while, I'd bring black garbage bags full of them, as well as more elaborate tortilla towels and *huipiles* (embroidered blouses), to community events and craft fairs. These efforts never garnered more than a few hundred dollars. In the parking lots of strip malls throughout the Pájaro and Salinas Valleys, seasonally unemployed or disabled farmworkers also peddled hand-beaded key chains, shell earrings, and tamales, going from parked car to parked car, doing whatever they could to earn cash to feed their families.

When a family member passes away, folks in farmworker families host impromptu car washes at gas stations and church parking lots to raise funds for a decent burial. At the area flea markets, there are stands that urge farmworkers to purchase life insurance so that their family members won't have to endure the shame of begging in the streets or getting a loan. Sometimes, these operations are fraudulent and exploit workers' fears.

FIGURE 15. Life insurance for farmworkers, for sale at a flea market. (Photo by the author.)

Aniceta died in March 2019 from a serious infection following surgery to adjust her pacemaker. Her family has asked me to help them sell the rest of their mother's *artesania*. Accompaniment as an act of politicized care carries on, even in the afterlife.

ACTIVISM AS CARE WORK

In accompanying *compañeros/as* in situations of distress, disease, and despair we can as anthropologists mobilize our skills as interpreters and navigators of languages and systems, and our time and energy, however constrained. As Melissa Checker observed during her fieldwork with African American environmental justice organizers in Georgia, "people don't always need an anthropologist," but often do need other things (2005, 193). The concrete effects of these relatively invisible and highly emotional and caring labors, as Whitney Duncan (2018) argues, are severely limited as far as effecting the structural changes that might have prevented and palliated Aniceta's illnesses, pain, and disease, or put off her early death, or perhaps avoided her family's need to leave Oaxaca in the first place.

I routinely feel useless. Very little changed for the better for Aniceta up to the day she died. Yet, being there, when I could, had profound effects that I still think are important. Aside from the empirical value of our presence through participant observation, accompaniment holds social and emotional value with political

FIGURE 16. Our Lady of the Live Oak Shrine at Pinto Lake, Watsonville, California, where im/migrant farmworkers come to pray for and mourn loved ones. (Photo by the author.)

and imaginative power. Armanda, Hortensia, Oralia, Aniceta and her children, and others taught me a lot about how impossible and unreasonable it is to examine all of these vulnerabilities, disparities, and injustices in isolation from one another. When it comes to health, all of these things combined have cumulative and cascading effects (Manderson and Warren 2016).

My involvement in Aniceta and her children's lives took on forms and feelings that I did not anticipate when I started my research. During one stay in the hospital in 2012, after her pacemaker gave her a repeated series of terrifying shocks, and weakened by her illness, a mixture of medications, and scary tests and procedures, Aniceta asked me, in a frail voice, to accompany her in yet another way. Would I be her *comadre* and her sons' *madrina*, or godmother, for their first communion?

Initially, I did not know what being a *madrina* entailed. I am an Anglo-descended non-Catholic woman who never had godparents. I envied the bonds and seemingly endless circles of aunts, uncles, and cousins characteristic of many im/migrant farmworker families. Mostly, *comadreando* and being a *madrina* has involved deep listening, both in person, over long phone conversations, and through WhatsApp chats, and the exchange of emojis, life advice, and long silences with teenage boys about really difficult things: depression, fear, anxiety, addiction, sexual assault, school woes, bullying, fights with significant others, troubling medical results, and the struggles and sufferings of family members back in Oaxaca.

I visit when I can, which is not as often as I would like. The three-hour drive to the Salinas Valley from Fresno takes its toll, as do the intense demands of my work at a public teaching university. There I manage heavy class loads and mentor students who themselves are from im/migrant farmworker backgrounds and are enduring their own multilayered life, family, health, environmental, and academic challenges.

The fictive kinships of *comadre* (literally, "co-mother," but also a close trusted friend) and *madrina* are ways that Mexican im/migrant communities continue to create spaces of emotional care and sometimes the mutual exchange of material aid and the sponsorship of parties, coming-of-age rituals, good food, and fun times that help folks get by. Aniceta and her youngest sons taught me how these relationships create connections of support and solidarity in situations of vulnerability, akin to what Carol Stack has described among African American folks as "strategies for survival in a community of severe economic deprivation" (1970, 28).

I know that my words, gifts, and visits do very little to change my *ahijados'* (godsons') material circumstances or their vulnerabilities as young Brown im/migrant men. Yet when anthropologists accompany, befriend, and become *compañeros/as*, or *comadres*, we participate in these still deeply political forms of care, a kind of care that does not always happen in the food banks, clinics, fields, nonprofit and social worker offices, and courtrooms I have described in this chapter. We "disrupt expected academic roles and statuses," Scheper-Hughes notes, "in the spirit of the Brazilian 'carnivalesque'" (1995, 420). Anthropologists have described their ethical responsibilities to and relationships with communities varyingly as "engagement," "anthropological citizenship," "the pact," "reciprocity," "pragmatic solidarity," and "collaboration" (Checker 2005; Dilger, Huschke, and Mattes 2015; Farmer 2004; F. V. Harrison 2010; Holmes 2013; Johnston 2010; Lamphere 2018). Each implies an obligation to contribute, give back, commit, or be present in the lives and well-being of researched communities. It merits a mindfulness about our intentions and how these are perceived and received in communities (Biruk 2017, 2; Huschke 2015:, 54, citing Deloria 1988, Medicine 2001, 289, and Tuhiwai Smith 1999).

Thus, the ways we reciprocate, or how and why we give back, engage, collaborate, make commitments, accompany, act, and react in solidarity matter a great deal, both emotionally and practically; yet anthropologists do not always write about this part of their work (Huschke 2015, 55–56; see also Hale 2001, and Heyman 2003). This includes erasure of the emotional and caring work entailed in getting involved in people's lives and struggles, the feelings and everyday acts that may lead up to participating in the more visible political and social movements that garner attention and respect as activism. While we may frame research as programmatic and activism as a series of grand and hypervisible gestures, the largely invisible affective work of caring about and for people in engaged research are also forms of activism. So, too, are the practices of learning and working toward change even when we fail (Salomón J. 2015, 188).

Accompaniment does not always have precise beginnings or endings or easily measurable or predictable outcomes. It consumes a lot of time and energy. It often left me feeling burned out, angry, and despondent. Those we accompany are also tired—of waiting in line, of being passed off from agency to agency and organization to organization, of feeling sick and tired (Hamer 2011). Emotions and engagement enrich and enliven fieldwork but also make it heartbreaking (Bourgois 2009, 2010; Checker 2005; Hale 2006; Holmes 2013; Nader 1972; Pine 2013; Saxton 2015b; Scheper-Hughes 1995).

Accompaniment is not always a pragmatic or practical endeavor. Still, our relationships to one another hold politically and socially transformative potential that stands in stark contrast to the antisocial politics of dehumanization and ecosocial abandonment characteristic of market-based medicine (Farmer 2004; Rylko-Bauer and Farmer 2002), agribusiness, state governance, and nonprofit and philanthropic work (Kohl-Arenas 2016; Wark and Ravéntos 2018). The intergenerational, interethnic, and transoccupational solidarities among teachers, students, farmworkers, and labor leaders that emerged during the campaigns against toxic soil fumigants present even more models for politicized care work—what I call ecosocial solidarities.

5 · ECOSOCIAL SOLIDARITIES
Teachers, Students, and Farmworker Families

Sometimes activism is loud and hypervisible. It feels raw and harsh in your throat and it makes your body tremble.

My first screaming match with a California state policy maker took place at the Watsonville Farmers' Market in May 2012. This is the time of year when strawberry and other fruit and vegetable farmers come back to direct sales following a six-month winter hiatus. I spotted him from a distance, enjoying a snack from a food truck. After shaking hands and making introductions with other activist friends, his smiling face shifted, turning stern, after he recognized me. To paraphrase, he said, "You're the one who wrote that article about me! I did not appreciate that! You have no idea of all the work I have put into this pesticide issue—engaging in dialogues, bringing people to the table. You should have come talk to me first before writing that!"

The op-ed that frustrated and apparently threatened him so much was something I had authored with collaborative feedback and encouragement from area anti–methyl iodide activists, including some of his supporters and constituents (Saxton 2012). In the piece I critiqued a newspaper article that quoted him reiterating the absolute economic necessity of soil fumigants. The article came out one day after Arysta LifeScience the patent holder and manufacturer of methyl iodide, voluntarily pulled its U.S. Environmental Protection Agency (U.S. EPA) permit for the pesticide, citing the product's low sales performance in California and the United States and the company's decision to pursue markets in other countries.

Too often, especially for chronically impoverished communities, jobs are framed as the ultimate solution to societal woes, from recessions and depressions to rising crime rates. I remain skeptical of this reasoning, noting that the quality and safety of jobs that come with economic growth policies and development plans are not often considered when deals with businesses promising jobs are brokered. I align with other scholars and activists who reject the harmful dichotomy of ecology and health versus jobs and economic growth. Instead they work from a "politics committed to *increasing the power and health of food chain workers*, and more broadly, *the communities within which they live*, by rejecting the tradeoff between food and jobs, which empowers working class people to shape the development of their communities" (Myers and Sbicca 2015, 17, emphasis added).

Over the previous two years, before becoming a state-level elected official, the policy maker had demonstrated consistent support for local grassroots antipesticide organizers in his region. At that time, teachers, youth, and labor and environmental activists had been urging city and county governments to submit letters of concern about the health and environmental risks posed by methyl iodide to then governor Jerry Brown. Brown had by then replaced former governor Arnold Schwarzenegger, whose administration approved methyl iodide in 2010 for agricultural use despite strong evidence of its toxicity and health hazards. To recap, methyl iodide is a highly volatile vaporized liquid applied to the soil to kill all pests, from rodents to nematodes, before crops are sewn or transplanted. It is a known carcinogen, miscarriage and birth defect inducer, respiratory irritant, neurotoxicant, and thyroid toxicant. Life scientists from the University of California–Berkeley issued public comments stating that they only used it with the most extreme caution in their laboratories, if at all.

Merely one day after Arysta LifeScience pulled its highly touted product from the U.S. market in March 2012 due to "poor market performance," the policy maker changed course. He urged that farmers be allowed to use other toxic soil fumigants lest agribusinesses suffer catastrophic losses. His argument and sudden shift in framing were unsurprising but still upsetting to those who had worked with him previously, and especially to those with sustained involvement in contentious and ongoing struggles for environmental health and justice.

I felt a seething anger amid the intensity of our encounter at the farmers' market. I got closer to his face, towering over him with my height. My body clenched. I screamed at the top of my lungs. Why was he so inflamed by what I wrote, which simply made an argument for more caution and skepticism around the role of pesticides in creating jobs? I don't remember exactly what I said during the heated encounter, which was hostile, admittedly, from both sides. I do remember how I felt after standing up to him with my body and my words. He shook his head and wagged his finger at me, saying, "You shouldn't have written that!" I shook and sobbed inconsolably and loudly on the curb for a while, as adrenaline and emotions seeped through my body in a stress response. Later I learned that my tears had embarrassed him. It would not be the last time he faced anger and criticism from the im/migrant and farmworking community he represented.

Activist and farmers' market friends who witnessed the argument comforted me with words, pats, and hugs. Later, in private, I talked it out with others. Some thought I had reacted too strongly, *too emotionally*; maybe the policy maker should have been given more of a chance to explain his reasoning. With my out-burst, I had burned a bridge: I would probably never have a seat at his negotiat-ing table ever again. As time went on, and the more I saw this pattern repeated with other policy makers who abruptly changed course when deliberating matters of public and environmental health, losing the possibility of a work-ing relationship with him and select other elected officials concerned me less and less.

Others who had tangled with him before knew him to be untrustworthy for his ambitions: someone who would make statements, take actions, and foster alliances by any means necessary to advance his career. Another activist friend, Patricia, described her own hostile encounter with a medical doctor whose treat-ment plan was causing her family member harm. She remembers how her tiny body and loud voice trembled and wept for long after the encounter as she, too, came down from the surge of feeling triggered by standing up to this powerful person.

Since this time I have had several other much more benign and/or engaging conversations with politicians, decision makers, and others with significant power in the agricultural industry who make the same pleas about pesticides and jobs. These talks don't always yield results as far as activist goals are concerned, and even more rarely change hearts and minds. They do create spaces where listening to the ideas of those who hotly disagree with environmental justice concerns can help inform strategies, even when the opposing side asserts that activists are unrealistic and impatient in their demands and desires for a global food system without toxic pesticides and other harms.

Whether through screams or even-toned statements and scribblings of scien-tific and ethnographic evidence, I have never regretted my choice to align myself firmly with movements seeking to challenge harm industries. Taking a stand is an important part of anti-oppression work. The activism for farmworker health and justice that I've been involved with has sometimes been explicit and out-spoken, such as through lively protests and rallies featuring homemade noise-makers and rhythmic chants. Heightened energies and emotions are also seen in activism's popular theatrical performances—for instance, when Las Lideres Campesinas, a statewide farmworker women's group, puts on improvisational plays for fellow community members that reenact how pesticide exposure, sex-ual harassment, domestic violence, racism, sexism, and anti-immigrant policies affect their everyday lives at work and at home. Petitions drafted by community groups and testimonials offered at public hearings also amplified and articulated community concerns in efforts to convince others.

Gladys, whose bold action is described below, gave up a college scholarship to make a statement about the injustices that farmworkers face on both sides of the border. A farmworker solidarity nonprofit forwent funding from agribusi-nesses to maintain the integrity of its work and relationships. The folks I met in

San Quintín, Baja California, México, in March 2016 sacrificed days of work to fight for fairer compensation, human and legal rights, and dignified and safe working conditions. Former farmworker Luz reached out to me for advice and insights when his older brother, Gerardo, had been diagnosed with Parkinson's disease that they both suspected had something to do with chronic pesticide exposure (as will be discussed later in this chapter). Pájaro Valley students did extra homework after school, and teachers and labor leaders added to their already endless schedules, to demand nontoxic and safe places to work, study, play, and come home to.

Why do people participate in such endeavors, especially when it sometimes feels as if the odds are not in their favor? Why do people persevere when companies and political figures with so much more power and wealth find ways to appropriate, subvert, and dismiss activist demands or to subdue them through charity, threats, violence, or shame? What are our relationships and responsibilities to the people and communities we know, or to those who are far away but at the same time interconnected to us through food chains, multinational trade agreements, and im/migration? Why do people who come from different places and life experiences care about and for each other? Why do we participate in social and environmental justice work when we are really under no obligation to do so? What does care look like? What does it feel like?

I agree with activist anthropologist Emma Louise Backe: "When you embed yourself within a vulnerable community, their struggle becomes yours. As an ethnographer, objectivity is less important than maintaining a critical sensibility, even in a space of solidarity" (2017, e84). In this chapter, I argue that there is more to activism than what's audible and visible, and that its value goes beyond the formulaic expectation that putting effort and energy into something must generate results or a tangible outcome in order to be successful (see Orñelas 2019). Ethnographic strategies of critique help challenge norms and assumptions that if left unquestioned continue to perpetuate harms. Methods of mobilizing activism and critiques vary greatly. People's everyday and emotional relationships to one another, through kinships and empathetic acknowledgment of shared challenges across occupational, geographic, and social locations, produce powerful reactions to toxic injustices in farmworker communities binationally.

The activism described herein represents some of the many examples of intergenerational and transoccupational solidarity I observed and participated in with farmworkers, their children, public school teachers, and other community allies. Activism, engagement, and accompaniment are not always obvious and explicit, taking place through the routines of our ethnographic labors and human relationships and the everyday acts of care expressed within communities (Besteman 2015; Howe 2016; Williams 2015). Cymene Howe builds on the idea of "negative space" from the arts, wherein blankness and openness strengthen a creative piece. In activism, negative spaces include ordinary people doing everyday things

(2016, 162–163): sitting in a waiting room with someone, giving a ride, washing the dishes, doing homework with teenagers after school, navigating college tours and applications, participating in a sewing circle where women commiserate over their work and family struggles, and going to meetings with elected officials and lobbyists. These are some examples of the invisible, mundane, and sometimes boring and routine activities that I observed and/or engaged in and that contribute to health and environmental justice work in agricultural communities (see also Checker 2005; and Williams 1984, 2001).

Activism is also rooted in our emotions, lived experiences, and observations—things that are sometimes quieter and less visible but no less significant. Teachers worrying about their students' health and learning, and children thinking about the welfare of their farmworker parents and taking steps to defend and protect them, are examples of some of the affective labors that informed health and environmental justice organizing during campaigns against toxic soil fumigants and beyond. Like the auxiliary nurses in Honduras described by activist medical anthropologist Adrienne Pine, decades of teacher and student antipesticide activism reflects "reinforced . . . interlinkages" (2013, 144) between the health and welfare of different occupational, generational, and social classes of people who come into community together through public institutions like schools and hospitals or the social relations of families.

Over the span of three years, my version of multisited fieldwork (Marcus 1994) put me in number of social, emotional, and physical positions. I could be found visiting my friend Luz's brother, Gerardo, at their Watsonville home, thinking through the possible connections between pesticide exposure and Parkinson's disease. I often sprawled out on the floor of the teachers union office, helping to make information packets on methyl iodide for Watsonville City Council members and Monterey and Santa Cruz Counties' boards of supervisors. I attended agricultural occupational health and safety and pesticide development conferences to listen to how people in industry framed and/or dismissed environmental health issues. I spent hours with young activists on research translation and outreach efforts in their communities.

All of these people, places, and events helped me learn about farmworkers' ecosocial suffering in the Pájaro and Salinas Valleys. They also encouraged me, in different ways, to put my ethnographic labors to work beyond research and data collection. Through what follows, I want to think through the ways in which different experiences, emotions, understandings, and responses to methyl iodide and other forms of environmental and structural violence surfaced, collided, and coalesced, producing responses and reactions to the environmental harms of pesticides. These occurred across occupational groups and within farmworker families composed of different generations and levels of formal education. Inspired by their work, I too sought to make myself more immediately useful, as a researcher and *compañera*, through a mixture of screams, observant silences, and participatory reactions (Maskens and Blanes 2013, 266).

STRAWBERRIES FOR SCHOLARSHIPS?

One day, a nonprofit organization I collaborated with during my fieldwork received a brochure in the mail from the California Strawberry Commission, the primary trade and advocacy group for the industry described in chapters 2 and 3. The glossy materials highlighted a scholarship program. Any college-bound student whose parent(s) had worked two or more consecutive seasons in the strawberry fields could apply for a one-time two-thousand-dollar award. In a different mailing, this time from the corporate philanthropy department of a major grower-shipper, an outreach worker inquired how they could help support the work of the organization, encouraging them to apply for a ten-thousand-dollar grant from the company's foundation.

The executive director of the nonprofit smirked. The idea of accepting funding from or partnering with businesses that employed the impoverished, injured, and disenfranchised farmworkers she had worked years to build trust with and support made her feel skeptical and uneasy. She forwarded the scholarship announcements to the college-bound youth in her networks, but in general ignored and distrusted these kinds of offers. She wanted nothing more than for all children from im/migrant farmworker families to have positive and enriching educational experiences and equitable access to higher education. This was in line with farmworker parents' aspirations too. Time and time again, farmworkers reiterated how they did not want their children to suffer in the fields like they had to.

Accepting charitable donations from agribusinesses felt too much at odds with the organization's mission to address and change the conditions that created social and health inequities for farmworkers in the first place. As has been noted in previous chapters, many of the layered disparities and vulnerabilities farmworkers endure are directly connected to agribusinesses' harmful labor, international trade, and toxic crop production practices. The organization's approach felt more holistic—syndemic, even—as it linked policy advocacy, accompaniment, material and mutual aid, and outspoken activism to address the multiple and intersecting injustices, exposures, and inequalities that disrupted farmworker families' health and well-being. The organization focused on and allied with other groups addressing environmental health harms, violence at and on either side of the Mexican-U.S. border, immigration injustices, and disparate access to safe and affordable housing, education, and health care.

We were not alone in our critiques of or reactions toward agribusinesses' charitable corporate social responsibility and philanthropic endeavors. Gladys

Morales, a student at California State University–Long Beach, hails from a Mexican im/migrant farmworker family. Her parents labor seasonally in the strawberry fields of Oxnard, just north of Los Angeles, on California's southern Central Coast.

To help fund her education Gladys applied for and received a California Strawberry Commission scholarship. At the awards ceremony, cosponsored by the commission and berry brand Driscoll's, she graciously thanked her family, friends, and teachers for their support. As she stood behind the podium, speaking into a microphone, she abruptly announced, however, that she would not be accepting the award: "I want to say to the California Strawberry Commission, 'thank you but NO thank you!' I want to say that dignity and self-respect has no monetary value and that I reject this scholarship because I also stand united with the farmworkers. Viva la lucha estudiantil y obrera! [Long live the student and worker struggle!]" (CTPP 2015). In her brief and bold speech, Gladys highlighted the contradictions of a scholarship sponsored by an industry that employed and exploited her people. Her family and millions of others had been involuntarily uprooted from homelands in Central America and México. There, agricultural livelihoods have become less and less viable due to trade policies privileging market, land, and water access for foreign agribusinesses and food imports over the rights and economic autonomy of Mexican farmers and workers (Daria 2019; Gálvez 2018; Otero 2018; Zlolniski 2019). Few farmworker families could save up enough to get through the lean months when the strawberries and other crops went dormant, let alone put aside funds for their children's college educations. How could a scholarship in the amount of a few thousand dollars given to about a dozen students each year make up for generations of suffering and sacrifice endured by im/migrant farmworker families' like Gladys's?

Gladys's statement also reacted to recent global events. In 2015, just five hours south of San Diego in the San Quintín Valley of Baja California Norte, thousands of Indigenous migrant farmworkers went on strike. They had migrated to the region to work harvesting strawberries, cucumbers, tomatoes, snow peas, fresh cut flowers, and other crops grown for large companies and destined for export. Together they stopped work for weeks, blocking highways, marching and hosting demonstrations in protest of poverty wages and hostile and dangerous working and living conditions in and around the fields. This initiated the first independent farmworker union in México, the Sindicato Independiente de Jornaleros Agrícolas, or SINJA (Independent Farmworkers Union) (Daria 2019). SINJA asked compatriots and allies in the United States to support it by hosting their own protests and demonstrations and educating consumers about the plights of farmworkers in México through outreach and a boycott of Driscoll's products.

In 2016, I traveled with my Fresno State University student Elio Santos to San Quintín to participate in a multiday march to Tijuana at the Mexican-U.S. border wall, commemorating the one-year anniversary of the strikes. Farmworkers organized the event with support from sister unions and human rights organizations in México and the United States. The route followed Highway 1 along the arid and mountainous Pacific coast, traversing a series of agricultural valleys covered by thousands of acres of fields. Some of these lands had previously been *ejidos,* farms managed cooperatively by peasant communities that produced shared food and income, and they dated back to the years following the Mexican Revolution. These swaths of industrial monoculture closely resembled the ones that had captured my attention six years earlier in the Pájaro and Salinas Valleys (as described in chapter 2).

Something distinct from California was the conspicuous signage proclaiming *precio justo* (fair trade) and *sin trabajo infantil* (made without child labor) plastered onto the gates and fences around fields throughout San Quintín. Identical placards emblazoned the backs of the white-painted converted school buses that transport farmworkers from *las colonias* (unincorporated settlements or neighborhoods surrounding larger cities and towns) to the fields: rides deducted from farmworkers' already paltry earnings of the equivalent of six to ten U.S. dollars per day. Chain-link enclosures surrounded vast acreages of U.S.-corporate operated strawberry fields, and these properties were heavily surveilled, sometimes with armed personnel. On the second day the marchers stopped at one of the gates to make a short informational film for folks following along on social media. They wanted to highlight the contradictions between product label claims of fairness, responsibility, and care, and farmworkers' lived realities. As we approached the perimeter of the property, a pair of ununiformed guards hissed at marchers and nearly hit some of us with a pickup truck.

Grower-shipper companies with operations in México, including BerryMex (a harvest management affiliate of Driscoll's) and Andrew Williamson and Company (which supplies many major grocers), have to date refused to honor, let alone acknowledge, SINJA's ongoing demands, as have the Mexican state and federal governments. Instead they have appropriated the language and models of fair trade, sweat- and child-labor-free movements to create private certification labeling schemes. Stickers pegged to these brand name cartons of strawberries at grocery stores proclaim, "responsibly grown, farmworker certified" or "fair trade." Not all fair trade or worker-certified labels are independently verified (Brown and Getz 2008; Guthman 2008). In reality, many represent a form of what labor anthropologist James Daria calls "fair washing" (2019). Instead of being organized and monitored by diverse groups of farmworkers, companies like Driscoll's and Andrew Williamson work with select high-profile nongovernmental organizations like the Equitable Food Initiative, the Mexican government, and what the farmworkers in San Quintín call the *sindicatos charros* (corporate and state corrupt unions) to set certification and monitoring standards and procedures. The collection of reputable affiliates behind the labels makes the seemingly benevolent

FIGURE 17. Farmworkers marching near San Quintín, Baja California Norte, México. (Photo by the author.)

efforts of grower-shipper companies appear legitimate and trustworthy. Participating agribusinesses profit from conscious consumers who attempt to make purchases that support their social and environmental justice values. All the while, the ongoing demands of independently organized workers are dismissed and ignored; branded as divisive, illegitimate, and unreasonable; and violently quelled through retaliation against dissenting or questioning farmworkers.

Farmworkers participating in the 2015 strikes and 2016 march were mostly Indigenous people hailing from communities in the Mexican states of Chiapas, Guerrero, Oaxaca, and Veracruz. Along the march route, I asked some of them if conditions in the fields or their lives had changed significantly since the fair trade and child-labor-free signs went up. Many shook their heads, scoffing in disbelief. "¿Precio justo? ¿Para quién?" (Fair trade? For whom?) I also marched alongside many workers under the age of sixteen who had forgone school to work in the fields to support themselves and/or their parents and younger siblings. Some of them had migrated alone. For youth who came with their families, their farm-working parents could not afford the bus fares needed to get their kids to the closest high school, two hours away. The labor of all able-bodied family members, young and old, was needed to survive. Rents on the tiny pallet-wood and cardboard shacks with plastic tarp roofs or the windowless concrete block homes with hand-dug latrine toilets; increasingly expensive food staples, toilet

FIGURE 18. Signs of "fair washing" on a fence. (Photo by the author.)

paper, and soap; and the municipally delivered rations of salty and toxically contaminated water were all beyond the salaries of the *jornaleros/as* (seasonal farmworkers).

Farmworkers and their supporters marched in March 2016 because they still did not have a *contrato colectivo* (collective union contract), a document sent to agribusinesses reiterating farmworkers' legal rights and demands for dignified lives and work. The *contrato* features statements about farmworkers' human rights to be paid fairly for their work at hourly and piece rates; to have workers' seniority respected when choosing people for promotions; to labor in workplaces free of hostility, racism, and sexual harassment and assault; and to be covered by legally required protections, including health care coverage through México's social security system, the Instituto Mexicano de Seguro Social and paid holidays (Daria 2019; Radio Bilingüe 2015). Some of these things are supposed to be guaranteed under the Mexican Constitution.

Other efforts in support of farmworkers in San Quintín included a boycott targeting Driscoll's brand products. Throughout the years 2015–2017, People hosted demonstrations inside and outside major grocery store chains like Costco, Safeway, and Whole Foods Market. With picket signs held high, protestors filed into stores, surrounding displays of berries and chanting, "No more blood berries!" Demonstrators plastered alternative labels onto berry clamshells that highlighted the hostile working conditions farmworkers in México endured. In grocery store parking lots, people handed out flyers about the collective contract and the boycott even after security guards asked them to leave. In the Pájaro Valley, activists protested outside Driscoll's corporate headquarters and dropped boycott banners off the sides of buildings downtown during the Watsonville Strawberry Festival.

Another demonstration took place at a Watsonville grocery store catering to Latinx residents that carried Driscoll's berries. We shared information in Spanish with farmworkers about what their *paisanos* (compatriots) in México were trying to do. Some people ignored us; others nodded in familiarity as they moved on with their errands. Others grew excited, using the moment to share their own stories of labor exploitation and hardship as im/migrant farmworkers in California. Some nonprofit organizations wrote letters urging organic grocers to stop carrying Driscoll's products. Two stores complied; the others argued that they either had no control over the supply chain or that it would be logistically infeasible to meet consumer demand for fresh fruit without buying from Driscoll's.

In addition to highlighting the limits of charity and philanthropy as models of care in farmworker communities, Gladys's speech astutely linked the lives of her im/migrant farmworker family to the struggles of Indigenous laborers harvesting strawberries under the same labels on the other side of the Mexican-U.S. border. She also aligned her arguments about labor struggles with the inequities many college students hailing from farmworker and other low-wage communities face. As the first in their families to attend college, these students with im/migrant parents must often piece together funding for school and living expenses while at the same time supporting their binational families. Many are managing full-time school responsibilities, working two to three jobs to make ends meet, and attending to their commitments to care for younger children, aging parents, and elders. This is part of what makes their activism so powerful: it defies the

FIGURE 19. Protesting Driscoll's outside a Costco store in Fresno, California. (Photo by the author.)

expectation that such groups are "insignificant" (Richter 2017), incapable of organizing and lacking agency.

SCHOOL DAYS

We hit the road at six in the morning, stopping only for breakfast burritos and coffee on the long drive to Sacramento. The route northward took us through varied industrial landscapes: strawberry fields, suburban track homes, state and federal prisons, and oil refineries. Diana, a high school teacher and teachers union organizer; Pepe, one of her former students and the son of farmworker parents; and I, their anthropologist, set out to meet two leaders from the statewide teachers union, the California Federation of Teachers (CFT).

Rural teachers from throughout California, in collaboration with groups like Californians for Pesticide Reform, Pesticide Action Network, and Pesticide Watch, had recently brought their concerns about routine pesticide exposure at rural school campuses abutting farm fields to the CFT's 2011 annual meeting. Several teachers and I had collectively drafted a resolution to present to other CFT delegates. We framed pesticide drift as a student and workplace safety issue, in addition to highlighting the science on soil fumigants' effects on children's health and learning. The resolution read in part,

Whereas, many schools in our state, are adjacent to fields where pesticides are in use; and

Whereas, our students are often forced to walk in proximity to treated fields to reach public transportation and to walk home; and

Whereas, despite efforts by the [California] EPA to instate drift mitigation measures, based on our experiences as educators and residents, we feel that buffer zones and other strategies are ineffective at preventing exposure to these life threatening substances; and

Whereas, children are significantly more vulnerable to pesticide poisoning than adults; and

Whereas, past and present studies indicate that pre-and post-natal low-dose exposure to pesticides may be a contributing factor in the development of learning and developmental disabilities in children; and

Whereas, a significant number of our students' parents work and live near the fields, exposing themselves and secondarily their children to hazardous chemicals, sacrificing their own health in order to make a living and support the well-being of their families; and . . .

Whereas, the overall potential negative impacts of living and going to school near ranches, farms and fields on which methyl iodide is used for pest management—even with the legally required protections, application procedures, and buffer zones in place—pose unconscionable risks to the health and well-being of our children, their families, educators and school employees.

Therefore, be it resolved, that the California Federation of Teachers request that further independent research be required by the state of California on the health and environmental impacts of methyl iodide, as it pertains to its use in commercial agriculture; and

Be it further resolved, that the CFT request that the state of California withdraw approval of methyl iodide for use in agricultural production until this research is completed, published, peer reviewed, and made publicly available in English and Spanish; and . . .

Be it further resolved, that the CFT call on CalSTRS [the California State Teachers Retirement System] to immediately divest from Permira until the private equity group sheds its investment in Arysta LifeScience or directs that firm to stop manufacturing methyl iodide for agricultural use; and

Be it finally resolved, this action is not only critical because it directly protects teachers, school employees, schoolchildren and their communities from a hazardous chemical, but also protects teachers' retirements funds tied to Permira/Arysta from risk of future litigation resulting from the use of this dangerous pesticide. (California Federation of Teachers, 2011, Resolution 1)

Teachers also presented a proposal to divest their retirement funds from Permira.[1] Both the resolution on methyl iodide and the retirement fund divestment proposal passed with widespread teacher support and got the attention of some policy makers, journalists, and other labor groups.

Shortly after this declaration of support, concern and solidarity from California teachers, statewide CFT leaders received an email from an agribusiness lobbyist, Mr. Soares (the name is a pseudonym), who proposed a meeting with the teachers union to discuss its recent activism against methyl iodide. The CFT solicited participation from members of the Pájaro Valley Federation of Teachers (PVFT) who had been the most active and involved in the methyl iodide campaign. Absolutely no activists from antipesticide nongovernmental organizations like Pesticide Action Network or Pesticide Watch would be permitted to attend, per the lobbyist's conditions.

Diana, Pepe, and I met with statewide CFT leaders outside a tall tinted-glass building in downtown Sacramento. We entered a glass fish-bowl style conference room in an office space where the lobbyist, three growers, and four representatives from various grower-shipper lobby organizations, including the California Strawberry Commission, greeted us. Firm handshakes and polite eye-contact ensued. Mr. Soares, the lobbyist, urged us to help ourselves to trays of cookies, soda and bottled water, but we all felt too anxious to eat.

Mr. Martin, Mr. Gomez, and Mr. Yang (the names are pseudonyms) each shared their positive experiences using methyl iodide. Mr. Martin and Mr. Gomez, both strawberry growers, had participated in research trials of methyl iodide with the University of California Cooperative Extension and Arysta. The use of demonstration plots to test and highlight the effectiveness of pesticides has a long history in California and elsewhere (Henke 2008) and is as much about research as it is about generating buzz around a new product. Mr. Yang was among the first of only six farmers statewide to apply methyl iodide commercially, using a free sample from Arysta to pretreat one of his vegetable fields. He took great pride in the fact that he grew culturally familiar and affordable food for the Central Valley's large Southeast Asian community.

Mr. Soares facilitated the conversation. He saw it as an opportunity to educate teachers, students, and union leadership about the challenges facing California agriculture. In his words, more and more "crop protection tools"—code for pesticides, including soil fumigants—were being banned or restricted to the point that their use proved too impractical or expensive. The three growers and the grower-shipper lobbyists spent a significant amount of time contesting the "realities" that had been outlined in the CFT resolution text.

I had participated in the resolution drafting process with teachers and PVFT leadership. We reflected on the fact that recent cuts to school bus transportation statewide meant that many more students in rural districts walked considerable distances to and from school each day. In the Pájaro Valley, this meant many students passed by or through recently fumigated or sprayed fields. I often observed this myself from the climate-controlled safety of my car: teens, farmworkers, and homeless people walked or biked along the roadsides.

In contrast, Mr. Gomez insisted that in his thirty years of farming he had never seen students walking in or near his fields. Neither he nor any of his workers had experienced any of the purported health effects of soil fumigants listed in the resolution and documented by scientists. This, of course, assumed that workers would be willing to report a work-related illness or exposure in the first place or would be able to link their symptoms to pesticides with any degree of certainty. Several studies coupled with the thin pesticide poisoning and injury reporting data described in chapter 3 suggest that farmworkers are often reluctant to report injuries and illnesses sustained at work (Holmes 2013; Horton 2016a, 2016b; A. A. López 2007; Quandt et al. 2000; Saxton 2013; Saxton and Stuesse 2018; Snipes, Cooper, and Shipp 2017; Stuesse 2018).

For Mr. Gomez and the others, soil fumigants like methyl iodide allowed them to achieve high yields from their strawberry fields and to provide jobs for farmworkers. Gomez's operation counted on sixty workers, 90 percent of whom returned every season to continue working for him. "I spend one million [dollars] each year just on labor," he explained. "I'd have to pay a lot more if I didn't have crop protection tools, and I can't afford that." Dividing a million dollars between sixty workers before taxes would amount to an average pay of approximately $16,600 a year at the wage and piece rates of the time and would include higher-paid forepersons, counters, and supervisors. Certainly many of the farmworkers I met made significantly less than $16,600 a year, and many had earnings below the poverty level. Some forepersons, supervisors, pesticide applicators, irrigation specialists, and other employees could earn a bit more, especially if they had year-round employment.

It was difficult to ignore some of the shared vulnerabilities and sometimes strikingly similar histories between these growers and their farmworker employees. For instance, Mr. Yang immigrated to the U.S. as a refugee. He shared with us how his family farmed using swidden, or slash-and-burn, techniques wherein farmers rotate crops and fields, burning the past season's harvest so the soil can regenerate for future years. These methods, while effective and ecological in their own ways, have long been ostracized as primitive and environmentally destructive by colonial administrators and, later, social scientists, aid workers, and conservationists charged with modernizing farming practices and feeding hungry people in third world countries (see Scott 1998).

Mr. Yang gazed at me, the anthropologist in the room, as he described how in order to survive as a small farmer growing specialty vegetables he needed the latest technology and tools available, including methyl iodide. Without pesticides he would not have a successful family business to pass on to his children. He would not be able to provide familiar healthy foods for his Southeast Asian neighbors. His personal history reminded me of the struggles faced by farmers in Central America and México, many of whom also feel a lot of compounding social, political, ecological, and economic pressures to farm with synthetic chemical pesticides, fertilizers, and genetically modified seeds (Dowdall and Klotz 2014; A. A. López 2007). At the same time, I could not so easily equate the challenges and complaints of the growers and grower-shipper lobbyists with those endured by farmworkers, their families, and teacher and community allies. The growers professed their environmental and social responsibility, framing their methods as an investment in the sustainability of agriculture and the health of farmworkers, neighbors, consumers, and the environment. Mr. Martin, for example, talked about his efforts to shift some of his production to organic methods. At the commercial scale, this required him to buy tons of expensive imported fish meal fertilizer that left a high carbon footprint. Empathy and a willingness to change only went as far as the constraints of the market.

Mr. Soares had convened the meeting as a way to humanize the growers and to foster greater empathy and understanding for the economic necessity of crop protection tools among the activists. It was an opportunity for them to share their experiences using soil fumigants and to argue face-to-face, not via picket lines and back-and-forth op-eds, as to why they needed methyl iodide and other "crop protection tools." The meeting, couched as an educational opportunity, attempted (unsuccessfully) to discourage teachers, students, the teachers union, and others from taking further action against methyl iodide. "What do we need to do to get you to stop?" asked Mr. Soares. I do not know what he had in mind as an offer: Scholarships? Funding for schools? The high school teacher Diana, while sympathetic to the stories of the growers, indicated that she and fellow activists would not stop organizing until they could be assured of student, teacher, and farmworker safety and well-being. Even after methyl iodide was pulled off the market by Arysta in March 2012, Pájaro and Salinas Valley teachers have continued their efforts to address ongoing health concerns about soil fumigants and other toxic pesticides still in use on farms and at or near school sites and residences.

In the end, we agreed to disagree. The meeting harbored no significant results in terms of any immediate wins for the movement, but this does not mean it was a waste of time. I respected and learned from the teachers' and students' refusal to give up in the face of more powerful interests' pleas and insistences, accompanied by cookies and soda pop. I also learned from the farmers and empathized with their challenges. In similar situations and contexts, anthropologists may be asked to take on the roles of brokers, negotiators, and mediators. At this meeting, teacher and student activists asked me to be a notetaker, and what Diana called "our brain": the one who would record and process the claims being made by growers and industry representatives and to translate that back into useful analyses and counterarguments that would help strengthen and sustain activist efforts. Some readers may find my explicit partiality off-putting or, even worse, biased. Social scientists witnessing or participating in community-based conflicts may be asked to carefully consider each side of an argument or debate and to give each side equal consideration. This is not always easy or ethical to do (Backe 2017) when different parties do not share the same levels of power or influence.

The distribution of the energies that go into toxic versus nontoxic or less toxic and worker-harming versus worker-friendly and fair production methods remains tilted toward the former. The burden of change to end if not reduce the disposability of labor is placed upon the most vulnerable folks in the debate. Meanwhile, farmworkers, their children, teachers, and rural neighbors live with the real and enduring challenges of earning and living on top of caring for their sick and disabled parents, siblings, and other relatives and contemplating uncertain futures. The ecosocial solidarities that formed between different communities linked to farmworkers as kin, friends, and allies provide alternatives to the uneven negotiations, sketchy alliances, and unfair compromises that grower-shipper companies angle for: from the eighth floor of that Sacramento skyscraper to the glossy pamphlets offering scholarships.

PESTICIDES, *HERMANOS*, AND *PAISANOS*: GERARDO'S STORY

In 2012, I sat at a small kitchen table turned makeshift desk in the Airstream trailer that served as my fieldwork "home." I had been struggling with the transition from fieldwork and activism to full-time dissertation writing. The pull from the community to keep contributing to various environmental justice campaigns conflicted with my needs to step back, articulate my findings on the pages, and finish my doctorate. It proved to be a very emotional process. Taking a teaching job at a local state university and turning off my phone for hours at a time helped somewhat.

But while my phone was still on, I received a text message from Luz, a former strawberry farmworker who had migrated with his older brother, Gerardo, from the Mexican state of Jalisco in the 1970s following their father's death. Luz wanted to know more about pesticides used in the strawberry fields in the 1980s, because Gerardo had recently been diagnosed with Parkinson's disease. I made time to visit, accompanying Luz as we walked up a home-crafted wooden ramp into a small one-story home where Gerardo, his wife, Claudia, and their twelve-year-old granddaughter Mariana lived.

Both brothers wanted to know more about the possible relationships between Gerardo's pesticide exposure, Parkinson's disease, and the pancreatic cancer that had claimed the lives of six coworkers. Among them were Gerardo's former employer, José Hidalgo—a *bracero* (im/migrant worker) turned strawberry sharecropper who rented land and grew strawberries under contract for a larger grower-shipper company. Hidalgo's two sons continued managing the farm after he died of pancreatic cancer in the 1990s.

In the 1980s Gerardo worked for Hidalgo, picking berries and applying pesticides with a crew of about ten other men. All, including Hidalgo, hailed from the same small *rancho* (village) in Jalisco. They pulled hoses, which was a method of pesticide application that is no longer used in California. One worker drove the tractor carrying the tank of chemicals (usually mixed manually by another worker), while others followed behind, heaving the long spray hose from row to row. This awkward, strenuous, and sweaty work required a lot of bounding over strawberry rows up to eighteen inches tall and close contact with the chemicals.

The brothers asked me to use my research skills and networks to see if I could find out what, exactly, Gerardo might have been exposed to during his time as a sprayer and farmworker. The only clues I garnered from our discussion were the name of the farm, Gerardo's memory of signing a form acknowledging the rules governing time-use restrictions on some pesticides, and his recollections about the Mediterranean fruit fly crisis, which involved lots of extra spraying. He also remembered being routinely gassed in the face when ripping up tarps following soil fumigations. When describing this, his face contorted, and the pitch of his voice rose as he reenacted the burning sensation in his nose and eyes. His pesticide exposure involved multiple products and many different routes of exposure (skin, breathing, eyes, etc.) over the course of his life.

I wrote an email to an agricultural extension agent who explained to me that Gerardo was likely exposed to organophosphates while pulling hoses. Organophosphate pesticides interfere with an organisms' electrical or nerve signaling between the brain and the body, causing paralysis, nerve damage, enzyme imbalances, twitching, tremors, and/or death. Other pesticides widely used in the conventional strawberry industry in California are linked to pancreatic cancer (Bassil et al. 2007; Clary and Ritz 2003; Ji et al. 2001; Moses 1999), the disease that killed at least six of Gerardo's workmates and his boss, as well as other cancers. Several of the most common soil fumigants used in California over the last thirty years—including methyl bromide, chloropicrin, metam sodium, and 1-3-dichloropropene—are known carcinogens, neurotoxicants, reproductive toxicants, and endocrine disruptors (Pesticide Action Network, n.d.-a, n.d.-b).

Gerardo remembered waking up on a Sunday in the early 1980s, unable to move. He had just finished six days and sixty full hours pulling hose. His condition gradually worsened, but he continued to work until 2011 when, at the age of sixty-three, his tremors became uncontrollable. That year a neurologist diagnosed Gerardo with Parkinson's and sent him to a specialist who enrolled him in a clinical trial. This helped a great deal, as he was no longer so dependent on the medications, like dopamine, that Parkinson's patients take to slow disease progression; however, these drugs also make people feel sick.

Gerardo's migrant cohort and coworkers kept tabs on one another through phone calls, random encounters in town, or gossip and word of mouth. This is how he heard about the deaths from pancreatic cancer of the *compañeros* who pulled hoses with him long ago. Many others moved back to México after retiring from farmwork. It is possible that they fell sick or faced untimely and unnatural deaths, too.

I used my university library access to look up scientific journal articles about pesticides and Parkinson's disease (Brown et al. 2006; Furlong et al. 2015; Kang et al. 2005; Moretto and Colosio 2011) and was able to use my research skills and bilingual language abilities to help Luz and Gerardo address their questions. Their goal was not to sue their former employer or to make any sort of public outcry; as Mario (in chapter 3) found, such legal resources were limited in their ability to achieve justice. Instead it was about trusting their instincts even as Gerardo's doctors, when asked about links between pesticides and Parkinson's, insisted that his illness would only be exacerbated by anxiety and anger and that he try not to think about things that would cause him distress.[2] Instead it was about trying to explore their ideas and suspicions that had developed, quietly, over the course of thirty years.

Listening and following up not only on the questions I had for Luz and Gerardo, but for the questions they asked of me, could be considered a form of pragmatic solidarity: practical ways of supporting the people that we work with while at the same time using our work to push for larger structural-level changes (Farmer 2004). It made anthropology a more mutual act (Besteman 2015; Sanjek 2015; Williams 2015). But as Seth M. Holmes has observed, there is a "need for solidary to move beyond only the pragmatic, the practical, or programmatic" (2013, 191). How can the limits of pragmatic solidarity, negotiating, brokering, compromising, and navigating be transcended? How do we merge activist concerns with the labors of anthropology in ways that are not necessarily practical or convenient but are urgent? The stories of public school teachers, teachers union leaders, and youth involved in antipesticide organizing demonstrate some possibilities.

STUDENTS: INTERGENERATIONAL AND ECOSOCIAL SOLIDARITIES

The children and grandchildren of farmworkers find themselves caught between two worlds (Zavella 2011). They see how they are the direct beneficiaries of their parents' labors in the fields, however precarious or temporary a living these jobs provide. A common refrain is that although farmwork wages are paltry, they are far more than one could earn farming or working in rural west central México. Many farmworkers and their children dream of other, more autonomous career possibilities: becoming farmers themselves, opening food trucks or restaurants, or running small shops, aspirations not so dissimilar from those they held in México (see Gálvez 2018). Others hope that someday they might ascend the agricultural labor hierarchy and become forepersons, supervisors, *ponchadoras* (farmworkers, often women, who inspect the quality and double-check the piece rates harvests in the field) packing plant workers, or tractor or truck drivers.

Farmworkers' children, surrounded by agribusinesses, sometimes attend college to study plant science or agricultural business so that they can have year-

FIGURE 20. Intergenerational and ecosocial solidarities: farmworkers and their grandchildren at an anti–methyl iodide demonstration in Salinas, California, 2011. (Photo by the author.)

round stable incomes with health care coverage and other benefits not normally afforded to those who work in the fields. Many families want their children to do well in school so that they might earn enough money to buy a family house in México and/or the United States. In this home, their parents, grandparents, aunts and uncles, whose bodies and spirits are worn from decades of hard farm labor, can retire and live comfortably while their kids and grandkids support them and keep them company as they age.

The dominance of agribusiness over the socioeconomic, political, and physical landscapes and dreamscapes of the Pájaro and Salinas Valleys did not foreclose questions and critiques of pesticide-intensive production from farmworkers and their children. Children's health proved to be a critical source of inspiration for sustained action around soil fumigants. Youths' experiences with asthma, learning disabilities, developmental delays, and witnessing their parents' daily struggles and health challenges made them think critically about agricultural pesticides.

Seventeen-year-old high school student Gloria sat with her peers in Diana's high school classroom. The students created a club to work on the anti–methyl iodide campaign, earning hours toward their community service graduation requirement, though this was certainly not their only motivation for participating.

As the eldest daughter of im/migrant strawberry harvesters, Gloria expressed concern for her family's health. Her friend, Ofelia, sixteen years old and six months pregnant at the time, also joined the group. For weeks they worked together with Diana and I to design bilingual informational materials. They shared them at one of their uncle's *loncheras* (taco trucks) where farmworkers congregated during lunch breaks. There Gloria and friends talked with family members and *paisanos* about the potential dangers of methyl iodide. The youth also developed a presentation of their findings and concerns for the Pájaro Valley Unified School District's school board and other city and county leaders, agencies, and organizations.

At a school board meeting in late November 2010, teachers, students from the migrant program, Watsonville Brown Beret members, and parents showed up holding signs that read "Protect Students' Health," "We Are Not Lab Rats," "Do Not Experiment on Us," "We Don't Want Cancer in Watsonville," "Moms Don't Want to Worry about Children," "Methyl Iodide Is Bad for Us!" and "We Want Clean Air." These slogans expressed very sophisticated and intimate understandings of the political ecology of pesticides, their suspected effects on the body, and a strident critique of the pesticide evaluation and approval processes used by the U.S. EPA and the California Department of Pesticide Regulation (CA DPR), which at times left too much lingering uncertainty.

At this school board meeting Ofelia took to the microphone, a hand rubbing her round belly, where her shirt was affixed with a sticker featuring a skull-shaped strawberry with crossbones, symbolizing poison and danger. She expressed concerns about her child's future, having learned about the potential for birth defects and developmental problems that could result from in utero exposure to methyl iodide and other pesticides. These worries were echoed by the CA DPR's Scientific Review Committee, a group of independent scientists convened to assess all available data on new pesticides. Committee members identified several problems with the methyl iodide review process. Their recommendations to exercise more caution with respect to known and potential or uncertain human and environmental harms posed by the fumigant were overlooked or ignored by the CA DPR in their final approval of methyl iodide in December 2010 (Froines et al. 2013, 1–2; Froines et al. 2010).

Gloria's mother, Rufina, expressed her own concerns during a visit I made to their home on the outskirts of Watsonville. Each room was painted in bright colors, reminiscent of the homes of México. One wall featured a large homemade shrine to La Virgen de Guadalupe, the much-adored patron saint of México. Gloria's younger siblings rode bicycles in the dirt driveway. On the clothesline hung freshly washed work clothes, stiffened dry by the breeze. Boots caked in red-stained mud from the fields stood in a neat row by the doorway. Their house, surrounded on all sides by miles and miles of strawberry fields, illustrated to me how difficult it was for folks to avoid exposure to toxic substances, even at home.

Every year applicators posted an official notice in English on their front door warning them of a pending soil fumigation or other pesticide application. Irrespective of the degrees of danger or the chemicals being applied, the family had nowhere else they could go. All they could do was keep the windows closed, stay indoors, and wait to do their laundry until the mandated reentry period—two days to two weeks after the application.[3] Rufina depended on Gloria to translate these notices. At the clinic where Rufina received prenatal care during her most recent pregnancy, she learned that exposure to pesticides can cause harm to babies. The risks included miscarriages, stillbirths, birth defects, and developmental disabilities.[4] She worried a lot about the health of her pregnant coworkers.

Gloria's father, Uriel, had worked on both conventional (pesticide-intensive) and organic berry farms and explained that he made less money when harvesting organic strawberries due to lower yields.[5] While he shared concerns with his daughter about the harmful effects of pesticides, he also worried that without soil fumigants to keep the soil "clean" and free of pests that harmed the berry plants he would not be able to support his family; banning or restricting soil fumigants could put him out of a job. With four kids to support, this thought scared Uriel, who defended his employers and their production practices. While Gloria's parents expressed pride in their daughter's accomplishments during her final year of high school, the immediate needs of feeding and sheltering their family came first.

As others have observed, farmworkers do not often have the privilege to choose one risk over another, be it the dangers of crossing the border or suffering heat exposure and illness in the fields (Holmes 2013; Horton 2016b). Workers like Uriel and other family members of students who met at the loncheras often explained that they couldn't (or didn't) worry about things like pesticides because, above all else, they needed to work. Yet when farmworkers are injured, disabled, or made ill by any number of the intersecting structural vulnerabilities of their jobs, im/migration, and home lives, their ability to work and support their families may be limited or cut short. This is apparent in the stories of Juan, Lilia, and Milagro (in chapter 1); Berta, Joel, and Mario (in chapter 3); Aniceta (in chapter 4); and Gerardo (in this chapter).

A more longitudinal view of im/migrant farmworkers' life courses suggests that the synergistic effects of different immigration, occupational, environmental, and household stresses and hazards, which can be slow to manifest (as in the case of Gerardo) or can happen in an instant (as with Aniceta), are not so easily reconciled by industry's claims for caring about workers or their welfare. Alayne Unterberger (2018) conducted longitudinal research with farmworkers in Florida and México and found that after succumbing to work injuries or illnesses, people are often forced to move back to their home communities, picking up even more precarious or unpaid work if they can, including childcare. Young, female, and healthier family members will often take on new paid jobs with long hours on top of caring for injured or chronically ill family members. While workers can't always afford to think about what *could* happen to them at work, that does not mean we should take the conditions of their labor as absolute givens or limit our ideas for social change work around what is possible in the immediate present.

After a while, Gloria and her classmates lost a lot of enthusiasm for the anti-methyl iodide work. They felt that too many of their farmworker family members and acquaintances were apathetic or indifferent toward pesticides. What difference could they possibly make without the support, participation, and interest of their own families and communities? Their demoralization intensified after December 1, 2010, when Governor Schwarzenegger approved the CA DPR's recommendation to register methyl iodide for agricultural use—against the cautions of its own Scientific Review Committee (Froines et al. 2010).

Interventions that have come out of public health partnerships with growers—such as getting companies and labor contractors to provide personal protective equipment to all workers and encouraging farmworkers to practice good hygiene to mitigate take-home pesticide residues on skin and on work clothing, and even proposals for pesticide-free buffer zones near sensitive sites pushed for by environmental justice groups—are important forms of harm reduction. But piecemeal efforts toward environmental and occupational health and safety in an industry riddled with harms are not good enough. Indeed, another value of the antipesticide activism I observed and participated in was its ability to instigate imaginative alternatives to agribusiness as a harm industry. Even if dreams of totally organic and agroecological production and farmworker-owned cooperative farms remained rare or elusive, a sense of indignity among teachers, students, and activists fueled their efforts to eliminate future generations' suffering.

Students set their sights on graduating and life after high school, fulfilling some of their and their parents' hopes and aspirations for the future. In the meantime, college students and community activists continued to work on antipesticide campaigns alongside public school teachers and teachers union organizers. They remain active to the present day.

TEACHERS: TRANSOCCUPATIONAL SOLIDARITIES

After walking into her rural elementary school classroom one September morning, Olga Harris (a pseudonym) could see plastic tarps spread across the fields adjacent to the school. She would later learn that this was evidence of a recent soil fumigation. As Harris (now retired) explained to me, "I remember grabbing on to a child's desk to break my fall . . . the dizziness just came out of nowhere." This incident caused Harris and her colleagues to ask questions about the connections between the appearance of the tarps and their symptoms. Their students suffered from chronic nosebleeds, headaches, and bouts of severe projectile vomiting that coincided with the appearance of the tarps and the applications of the soil fumigants methyl bromide and chloropicrin. Only chain-link fences separated classrooms, school parking lots, playgrounds, and residences from the fumigated fields, and only by mere feet.

Teachers approached the campaign against methyl iodide and other soil fumigants and pesticides from their deep concerns for their own health and that of their students and those students' farmworker parents. They observed and embodied pesticide exposure at school sites located close to agricultural fields, which includes many schools throughout the Pájaro and Salinas Valleys.[6] During my fieldwork I interacted with veteran teachers, mostly white women now in their sixties and seventies, who had worked for the Pájaro Valley Unified School District in the 1980s and 1990s. At that point the carcinogenic soil fumigant methyl bromide, which had been banned and scheduled for what became a very prolonged phasing out in accordance with the 1988 United Nations Montreal Protocol, was still being used (Gareau 2013).

In the 1980s and 1990s, teachers' embodied experiences with pesticide-related sicknesses and symptoms and their reactions to watching students and families suffer, both environmentally and socially, led to action. They joined some organic farmers, environmentalists, and regional and state-based community organizations to form a local working group. Using their educational and bilingual training they collaborated with the Environmental Working Group to collect data on the extent of pesticide drift around schools and rural households (Environmental Working Group 1997). They incorporated their findings into ecology, science and art curriculum and went door-to-door to share their concerns with students' families. These efforts culminated with the organization of a "sick-out" at Amesti Elementary School, wherein many parents kept their children home from school as an act of protest. This was intended to put financial pressure on the school district, which receives state funding partially based on student attendance.

At that time the Pájaro Unified School District Board of Trustees remained loyal to growers and the agricultural industry and warned teachers to stop scaring children and parents. Several teachers, including Carole and Julianne (pseudonyms), who also taught at Amesti, lost their jobs through firings and forced retirements. Julianne developed Parkinson's disease, and Carole died of throat cancer in 2012. Both suspected that their illnesses were related, at least in part, to years of school site and residential pesticide exposure. Carole did not receive a diagnosis until her cancer had metastasized to other parts of her body. That she waited so long to seek medical help was directly related to the loss of her teaching job, as she was chronically unemployed, sporadically homeless, and uninsured.

Teachers, students, and parents continued participating in antipesticide activism. During the time of my fieldwork, Diana told me about her first teaching assignment in Philadelphia. She worked as a reading specialist. Many of her students suffered from developmental delays resulting from chronic lead exposure. She remarked at how similar the learning difficulties were between her previous urban students and her current rural ones in the Pájaro Valley. Other anthropologists and environmental health scientists affiliated with the Center for the Health Assessment of Mothers and Children of Salinas have made similar findings with respect to pesticide exposure and learning disabilities and developmental delays in agricultural communities in California (De Long and Hollaway 2017; Dewan et al. 2013; Eskenazi et al. 2006, 2010; Gemmill et al. 2013; Holtcamp 2012; Marks et al. 2010; Rosas and Eskenazi 2008) and México (Guillette et al. 1998). Another teacher, Janet, ran the special education classroom at a school that was surrounded on all sides by strawberry fields. She noted how despite no significant changes in the overall student population, her classroom sizes continued to grow. She and other teachers struggled to serve the diverse needs of the learning disabled and developmentally challenged children of farmworkers; it was a struggle that was compounded by district-wide spending and hiring cuts, teacher layoffs, and outdated and deteriorating classroom facilities.

Fast-forward to 2016, when over steaming plates of *chiles rellenos* at a local taco shop, Olga Harris explained to me the relationship between the chronic health issues she'd been having and her decades of teaching next door to pesticide-intensive agricultural production:

> All of us who were involved [in antipesticide organizing] are very well educated. We have a feeling about the role of pesticides in all the things we've been observing with our own health and that of our students . . . we just don't know for sure. I have no family history of autoimmune problems. My colleague, who has terminal bone cancer, also has told me that she has no family history. But still, at this point, we can't know for sure, but that shouldn't stop us from acting upon what we do know . . . and there's a lot more research now than there was when we first started.

While we talked in the restaurant, one of Harris's former students greeted her with a huge hug, a wide grin, and "Remember me?" He started reminiscing fondly about all of the extracurricular activities they did in Harris's classes to learn more about the region's ecology, including the effects of pesticides. His wife, eight months pregnant, joined in. She, too, had attended school in the Pájaro Valley. As she disclosed her struggles with asthma, she suggested bluntly, "I blame it all on growing up here and attending McQuiddy School!" These kinds of conversations came up constantly when I lived in the Pájaro Valley, punctuating daily life.

ECOSOCIAL SOLIDARITIES: ANTHROPOLOGY AND BEYOND

One day, while I was working on this chapter during a visit to the Pájaro Valley Federation of Teachers union office, talk circulated about an upcoming Watsonville City Council hearing. Up for discussion would be the construction of a new hotel on the outskirts of Watsonville, coincidentally funded by agribusiness dollars from a wealthy almond grower from the Central Valley. It had been suggested by the hotel proprietors that a 250-foot buffer zone be instated in the fields that surrounded the hotel site so that pesticide drift would not stain the cars of visitors and tourists to the Monterey Bay area. Pájaro Valley agribusinesses were willing to work with the hotel.

Meanwhile, the same consideration regarding public and worker exposure to pesticides continued to be heavily debated and denied. In 1999 a civil suit involving parents, students, farmworkers, and environmental nongovernmental organizations argued that the disproportionate pesticide exposure endured by Pájaro Valley's predominantly Latinx student body constituted race-based discrimination under Title VI of the Civil Rights Act of 1964 (42 U.S.C. §2000d et seq), which "prohibits discrimination on the basis of race, color, or national origin in any program or activity that receives Federal funds or other Federal financial assistance." This includes public schools. Twelve years later, in 2011, the CA DPR and the California Environmental Protection Agency (CA EPA) acknowledged the legitimacy of students', parents', and teachers' claims but awarded no compensation. Instead they installed air quality monitoring equipment around schools on the Central Coast and in the Central Valley with the aim of studying the issue more closely.

In 2013 the original families and organizations refiled the civil suit at the federal level after the CA EPA dropped the original complaint without informing the teacher, student, and parent plaintiffs (Garcia, Newell, and Stano 2013). The preliminary results from the air monitors as reported by regulatory agencies suggested that pesticide levels in the air were at or well below legal or limits at the time of measurement (Vidrio et al. 2014). There are, however, a number of inconsistencies in how the CA DPR and CA EPA designed their monitoring study and how they interpreted the data. Their research to date has not accounted for the exposure experiences of past generations of teachers, students, and families, as they were not formally recorded or monitored. The methods used to establish pesticide exposure limits do not account for the lifetime effects of long-term low-level exposure.

Thus, a potential arena for solidarity among researchers would be to find funding and methods that better account for the synergistic and longitudinal effects of the toxics, stressors, traumas, and poverty that structurally vulnerable people endure. For health care and social service providers, educators, state workers, and others who have working or personal relationships with farmworkers, there are opportunities for solidarity by building trust and developing working groups and an infrastructure that document the effects of harm industries and strategize on how to respond to them.

The back-and-forth of legal and civil suits from environmental justice organizations, followed by calls for further studies on pesticides' health and environmental effects by regulatory agencies, then followed by inaction and thus calls for more research, is seemingly endless. Meanwhile, research and data continue to accumulate but are rarely used to put change into action. Protective policies may languish in legislative procedures. When laws are passed, there can be difficulties with enforcement and monitoring. Strategically, lawsuits are one of the few economic strategies that resource-poor communities have at their disposal—that is, when lawyers are willing to take them on. Lawsuits can have some financial impact on agencies and polluting companies that must spend millions of dollars to defend themselves in responding to these cases. This was one of the reasons Arysta LifeScience cited in its decision to voluntarily retract its CA DPR and U.S. EPA permits for methyl iodide (Guthman 2019; Guthman and Brown 2016b, 2017). Individual victims of exposure and other harms, however, rarely receive compensation,[7] as in the cases of Aniceta, Gerardo, and Mario. Additionally, a lot of money, energy, and political will is invested in delegitimizing research that threatens pesticide use or attempts to debunk agribusiness companies' claims that pesticides are safe (when applied following the product label's instructions) and pose no harm to consumers or the environment (Oreskes and Conway 2010). Industry and regulatory responses concerning studies and activist arguments have been overwhelmingly dismissive. Activists are often accused of being uninformed or uneducated, irrational and overly emotional, and farmworkers' symptoms and sicknesses are written off as merely psychosomatic.

The activist efforts of teachers in the Pájaro Valley may be considered tame in comparison to the radical responses of auxiliary nurses and resistance organizers in Honduras, who literally put their bodies in the line of fire, from bullets and tear gas, for their patients (Pine 2013). Certainly California teachers have rarely experienced direct physical violence in response to their antipesticide work. Still, teachers and teachers union organizers acknowledge the parallels between their own health, that of students and their farmworker parents, and the disturbing potential relationships between pesticide exposure and students' educational quality and outcomes. In these ways, akin to the actions of the auxiliary nurses in Honduras, teachers developed a culture of critical consciousness about how their work and health, and that of their students and their farmworker parents, is being undermined by the slow violence of pesticide applications near schools and residences and the inadequate legal protections.

Irrespective of the measurable *effectiveness* of their activism on the governance of pesticides, teachers' and students' *affective* expressions and efforts to imagine alternative futures (Burke and Shear 2014) for food, farming, and work deserve validation and support. In these ways, ecosocial solidarities among farmworkers, their children and grandchildren, public school teachers, and other community activists develop through both the *physical* and *emotional* dimensions of embodiment—wherein material and social disparities and discrimination, as well as consciousness, become incorporated into bodies and body politics (Smith-Nonini 2010) and social, moral, political, and ecological understandings of communities (Krieger 2010, 2001; Nixon 2011; Pine 2013). Ecosocial, intergenerational, and transoccupational solidarities have sustained decades of grassroots antipesticide activism in the Pájaro and Salinas Valleys that continues to the present day. In 2017, citing growing consumer demand for organics and pressure from teachers, students, and parents, the large berry grower-shipper Cal-Giant has decided to transition more acreage around schools, day-care centers, and other sensitive sites to organic production. This includes a fifteen-acre plot next to Ohlone Elementary School just outside Watsonville, one of the most historically contested sights in the antipesticide movement (Dennis 2017; Vanderhorst 2017).

One challenge, though, is the feeling that even with the small but significant concessions that agribusinesses sometimes make in support of worker health and well-being—for instance, when one pesticide is pulled from the market or more strictly regulated, or when any law protecting workers and the environment is passed—other challenges come up. The response cycle continues, or as the Mexican activist slogan goes, *la lucha sigue y sigue* (the struggle goes on and on). How to cope with this—what I call "triage mode" of environmental health and justice activism—is the focus of this book's Conclusion.

CONCLUSION
Activist Anthropology as Triage

I started writing this conclusion in California's Central Valley, a five-hundred-mile-long stretch of industrial agriculture dotted with ethnically and generationally diverse communities and dominated by super conservative blood red politics. The political transitions following the election of Donald Trump as president in 2016 have included a rapid succession of executive orders and explicit and casually dropped hostilities. These have targeted the poor, im/migrants, refugees, and people of color, LGBTQIA+ people, women, non-able-bodied persons, ethnic and religious groups, and specific groups of world travelers, among others. Anticipations of the rationing of social safety net and educational resources; the gutting of long-fought-for environmental, labor, and health care policies; poorly managed public health emergencies; and yet more endless wars loom ominously. All of this is especially distressing for the im/migrant farmworker families I have met and befriended over nearly ten years of engaged scholarship and activism.

In Fresno, California, where I now live and work, young people—including some of my students, who are im/migrants and the children of farmworkers themselves—have stopped going to school. The lines at food distribution centers in some of the most food insecure and impoverished communities, which are also communities where so much food is grown and harvested to feed others, grew sparse in 2017 (Bernstein et al. 2019), and then grew long again with the arrival of the COVID-19 pandemic in March 2020. In fall 2019, Trump announced the intention to foreclose pathways to legal residency for im/migrants who at any point had relied on any kind of federally funded support, including food, housing, and health care assistance. Known as the "public charge" rule, it is still being debated in the courts. Even though it has not yet been formally enacted, it sent many people into a panic. Many im/migrant families rushed to cancel their enrollments (and those of their citizen children) in programs like SNAP (Supplemental Nutrition Assistance Program) and Medicaid, cutting off already limited access to life-saving health treatments. Despite unchanged eligibility requirements in California, enrollments in state and locally managed safety-net programs have declined (Straut-Eppsteiner 2020). This is creating an avalanche of compounding problems during the coronavirus pandemic, generating new vulnerabilities and severely exacerbating preexisting ones.

Im/migrants exercise extreme vigilance whenever they leave home, fearful that a trip out the door could lead to encounters with immigration officials (IndyBay 2017). Families created emergency plans in the event that parents or guardians are separated from their young children. Some folks are afraid to go to work in the fields and packing plants, while others are self-deporting to preempt the familiar traumas of encounters with Immigration and Customs Enforcement (ICE; see Campbell 2019). On top of these fears, they face coronavirus infection and intense wildfire smoke in California and the Western States. The devil's fruit makes you damned if you do, damned if you don't.

Folks are lying low—lower than they already were before the election, having already witnessed previous generations of elected officials and law enforcement officers bolstering and holstering anti-immigrant policies. They remember the workplace and community raids of years past. They've experienced scapegoating and shaming for nationwide economic and social crises (Chávez 2013; Chomsky 2007; Molina 2006; Ngai 2014). Today, essential services and critical social supports are growing ever more limited for all, as have rights and protections for all kinds of so-called essential workers, including and especially farmworkers.

Vast amounts of the groundwater and drinking water in the Central Valley are contaminated, in large part from decades of industrial activity, including oil and gas extraction and heavy use of synthetic pesticides and fertilizers on the region's expansive farms. And there are dozens of farmworker communities in the Central Valley who have experienced periods with no access to water at all; industrial agriculture's intensive consumption of groundwater has sucked many residential wells dry and left remaining supplies saltier and more contaminated. At various points every year, air quality alerts warn that conditions outside are unsafe for vulnerable groups: children, the elderly, and people with respiratory conditions like asthma. On some days *all* people are advised to stay indoors.

The devastating effects of catastrophic wildfires that glare orange from our television, computer, and phone screens, and in some cases our backyards and farmwork sites, extend beyond the ignition points and burn zones. Heavy smoke, laden with toxic particulates and gases, travels throughout the state, and even internationally, on the winds. During the record-breaking California wildfires of the falls of 2018, 2019, and 2020, many farmworkers worked in even more dangerous conditions. In 2018 and 2019, Some agricultural employers actively attempted to prevent workers from receiving masks that visiting activist groups were seeking to distribute free of charge (Herrera 2018), so some activists took matters into their own hands (Morales and Lubin 2019).

For farmworkers up and down the California coast living and working in the twenty-first century, breathing in toxic smoke, job loss and reduced hours (despite an ongoing labor shortage), heightened risks of homelessness, and exclusion due to citizenship status from federal emergency relief efforts, have become ever more chronic and life threatening. Meanwhile, all the other problems described in *The Devil's Fruit* also remain constant.

These ethnographic vignettes parallel those observed in other parts of the country, from the Pacific Northwest to the Deep South, where im/migrant workers face—not new, but heightened—uncertainty, job insecurity, and ecosocial vulnerability (Kline 2019; Stuesse 2016). The situation feels dire and hopeless on many fronts. Why keep going when everything feels so out of control and hopeless, when small victories pale in comparison to the work that lies ahead? When smoke turns the sky gray makes the air poisonous for months on end? When we're living through a deadly global pandemic?

I have come to see activist research as not just a tool, theoretical framework, or methodology but a way of working and living as a human. In chapter 1, I discussed how the anthropological tradition of documenting folklore of cultures could be turned on its head to dispel widespread and dangerous mythologies about im/migrants and farmworkers. In chapter 2, I employed the ethnographic strategy of following the object: studying strawberries, their growing fields, and the people, processes, and toxic substances involved in their production. Defamiliarizing strawberries, which are often framed as healthy and sweet, generates critical and ethical questions about the monocultural and harmful practices used by industrial agribusiness and challenges some of the assumptions and logics of the global food system. In chapter 3, I discussed the scope of pesticides as a global health problem. Farmworkers' lived experiences of toxic layering and invisible harm, and the failure of policies and industry to address them, have resulted in syndemic patterns of ill health and ecosocial suffering throughout farmworkers' life courses and across generations. I have suggested that the critical medical anthropology concepts of syndemics and chronicities can expose and articulate these patterns and perhaps inspire more dynamic and ecosocially minded interventions, community organizing and reactions.

In chapter 4, I contemplated the limits and possibilities of different kinds of caring labors that I observed in farmworker communities. I came to see accompaniment as a form of care that can be mobilized in our research, participant observation, and interpersonal relationships. Accompaniment teaches us what is systemically wrong with health care and social systems that are failing im/migrant farmworkers in the United States. It also presents a more holistic and repoliticized way of doing caring work. In chapter 5, I framed some of these caring and emotional labors in environmental justice activism as ecosocial solidarities. The teenage children of farmworkers, and their teachers in the public school systems in the Pájaro and Salinas Valleys, acknowledge risks and concerns about pesticide exposure on and off the farm, at school sites and in houses adjacent to agricultural fields. Their care for their farmworker kin and neighbors defies arguments about the inevitable schisms between labor and environment and models the possibilities of transoccupational, interethnic, multilingual, and intergenerational alliances that have inspired my own involvement and sustained commitment to environmental health justice in rural communities.

It is one thing to study and theorize social problems and to postulate rational, financially feasible, and practical solutions, but we live in a time that is highly irrational, chaotic, and insecure. Throughout *The Devil's Fruit* I have argued that to mobilize research as a tool of change and challenge requires different kinds of relationships with our work and with the communities where we work. I see the emotional intimacy and caring of activism in research as a strength rather than an epistemological weakness. It makes our work more human and humane.

It is not always easy or rewarding, however. Activist research is more time-consuming and exhausting, both physically and emotionally. With that comes a range of disorienting feelings, including indignity, impatience, frustration, anxiety, depression, cynicism, fatigue, and the sensation of being overwhelmed and outnumbered. Farmworkers sometimes express it in a single word, *coraje*: the intense rage felt throughout the body and spirit (Horton 2016b). I know many activist researchers and other caring people who feel constrained, limited, and pulled in many different directions all at once.

Activism became a way to express and direct these energies and emotions. Making myself and my work useful to communities is hard but also heartening. I find joy in the connections and relationships I've made and sustained, in the bowls of pozole and in the fruits of our respective shared labors. Validating human experiences that are too often dismissed provides some relief. At the same time, I know that I and many others are overwhelmed and stretched thin by the unnecessary and unconscionable levels of suffering and institutional neglect we witness, and in some cases experience directly, from employers' or others' indifference. We may feel conflicted between doing what we need to do to stay employed (Chatterjee and Maira 2014; Johnston 2001a; McKenna 2004) and taking care of ourselves and our responsibility to mobilize our privileges to support environmental health and justice work. The concept of triage has helped me process these tensions.

DAYS IN THE TRIAGE MODE

On September 3, 2017, just as the fall semester ramped up at Fresno State University, then attorney general Jeff Sessions announced the pending termination of the Deferred Action for Childhood Arrivals (DACA) program. Authorized by an executive order during the administration of President Barack Obama in 2012, DACA granted certain young im/migrants who had come to the United States as children—those who met several strict qualifications, and who submitted their applications on time, along with a nearly five-hundred-dollar fee—the right to live, work, and study here. Every two years they would repeat the process (and pay the fee) to maintain their status.

That Friday evening leading into the Labor Day weekend, I felt numb. I remembered volunteering to help DACA applicants in Watsonville in 2012 complete the tremendous amounts of paperwork, and attending fundraisers to help young people cover the fees. I thought of all the folks I knew who had contributed, and the students, including my students, who were thriving.

The following Tuesday I stood in front of one of my anthropology classes. Tearful and trembling with a mix of rage and uncertainty, I explained what was happening. I passed out flyers from the Central Valley Rapid Response Network. Organized by Faith in the Valley, a nondenominational coalition, the network is a 911-like crisis hotline that provides prayerful and critical witnessing, comforting support, and strategic interventions whenever ICE, law enforcement agencies, or homegrown racists threaten the lives and well-being of members of the valley's im/migrant, LGBTQIA+, Muslim, Sikh, and nonwhite communities. Similar networks have been launched across the country. Volunteer responders learn how to document and disrupt detentions; how to accompany folks during court cases and immigration hearings; and how to mobilize resources, legal assistance, and other forms of support. Many of these groups depend on a small number of dedicated volunteers and are exploring ways to make their work more sustainable.

I didn't know what else to do in that moment. I'd never cried in front of students before. My students' commitment to stand by their undocumented peers and their families renewed my convictions. This tearful sharing, coincidentally, preceded a lecture on culture and power. Current events and tears made it more real and less abstract.

Later that month my student Martina (a pseudonym) and her parents had been brought into Fresno's regional ICE office for questioning. ICE officials dubiously claimed that her parents had failed to fill out new paperwork upon Trump's election, and therefore they would be subject to deportation proceedings. My attendance at several know-your-rights trainings and my involvement with community groups enabled me to challenge this information. I directed Martina and her family to the Rapid Response Network.

Even amid the emotional turmoil of the situation, Martina's goal was to stay enrolled for the semester. The terror and stress instigated by the ICE interrogation contributed to other, layered, problems. Her mother got sick and suffered complications, and Martina offered to donate her blood for a transfusion. Consequently, she fell ill herself. Throughout her life, she served as her parents' primary interpreter in all medical and legal matters.

It was a risk for Martina to disclose her DACA status to her professors, but she had to explain her absences. Eventually she stopped coming to class altogether. Over the phone she informed me that I had been the only professor who had offered her any help. This made me cry for weeks on end.

I grappled with the collective potential of my peers and our employer to do even more, yet they, too, face their own limitations and stresses. Days after Attorney General Sessions's announcement about the pending cancellation of DACA, the California State University Chancellor's Office sent out a statewide memo suggesting that waiting for a federal bipartisan resolution was the way forward. In the meantime, employees who were also DACA recipients, including faculty, staff, and work-study students, would lose their jobs once the executive order went into effect (White 2017, 2). University administrators throughout the California State University system would engage in dialogues with their state and federal representatives, some of whom were close allies of Trump and who endorsed strong anti-immigrant policies.

Since then, the ongoing boastful successions of declarations about student diversity and success, and about our responsibility as faculty to do whatever it takes to support our students, feels even emptier than ever. While DACA's cancellation has thankfully been indefinitely delayed by activists, savvy lawyers, and federal court judges, im/migrants—with and without citizenship status—continue to face threats. The Trump administration is neither the only administration nor the first to enact such restrictions. A long history precedes it, and a broader climate of xenophobia is creating similar problems for communities throughout the world.

This and other experiences and conversations with fellow activists, trustworthy colleagues, and so many incredible students at the university have deepened my thinking about what it means to be a professor, an anthropologist, an activist, a researcher, and a community member in times of unending, layered, and at times diabolical crises. I often feel frustrated by my limitations. More so, I am enraged at the failures of institutions in our society with significant resources at their disposal, including universities, to do more, to not be hampered in their efforts by the limits of the law and their imaginations, and fears about upsetting donors or key people in the business community.

I often think about the words of Catherine Mazak, a professor of English at the University of Puerto Rico–Mayagüez, one of the only facilities with access to power and water in the months immediately following Hurricane Maria: "What should the university be in a time of catastrophe? This: The center of service—a relevant and essential key to the survival, recovery and betterment of the land and the people it serves" (Mazak 2017). In the context of DACA, this means not waiting around for an unlikely bipartisan solution to U.S. immigration policy. We must remember the ghosts of anthropology's not-so-distant past and channel their examples to become a haunting force. Challenging and trying conditions have not prevented resistance, critique, and response from generations of communities. Activist anthropologists have a track record of using their knowledge, tools, perspectives, positions, and privileges "in an applied anthropology that engages, documents, promotes and supports cultural diversity, social justice and environmental sustainability" (Veteto and Lockyer 2015, 359) against all odds.

I have come to think of the suturing together of these different labors, levels of intervention and care, and accompanying emotions as a sort of *triage*, the need for which is only intensifying amid multilayered and intensified ecosocial vulnerabilities: toxic exposures, suffocating smoke, climate change, human and workers' rights roll backs, poverty, fascism, white supremacy, xenophobia, and racism. Triage is a commonplace practice in emergency rooms, disaster response, and crisis management efforts. From the French, *trier*, meaning "to sort," the concept was first applied in agriculture to describe the action of sorting goods (Iserson and Moskop 2007, 275). During the bloody Napoleonic Wars, the French doctor Baron Dominique-Jean Larrey developed triage as a method for rapidly assessing and treating patients based on the severity of wounds or illnesses, irrespective of one's social status or military position (Iserson and Moskop 2007, 276; Robertson-Steel 2006).

Despite these egalitarian sentiments, triage can also imply a "rationing" of care in times when supplies, medicine, funding, or doctors and nurses are in short supply (Iserson and Moskop 2007, 275). Different situations, from the daily grind of emergency rooms to the chaos of disasters, have different organizing logics that determine who will receive care, in what order, and of what quality, contingent on what material and human resources are available. Sometimes, in major crises or situations involving mass casualties, "some patients who could be saved may be allowed to die to save others" (Iserson and Moskop 2007, 276, 277). Amid the global economic crises of the 1970s, newspaper pieces used triage to describe decision making about the allocation of resources. From students in struggling schools to starving citizens of third world countries, aid and support would be prioritized for those deemed most likely to survive and thrive and denied to those who could not be helped or were less likely to survive.

In these ways, the sorting logics and practices of triage systems can take on pernicious and vicious forms that intersect with other hierarchies, values, and vulnerabilities. For example, Barbara Andersen (2014) and Vinh-Kim Nguyen (2010) have observed in the context of HIV/AIDS clinics in Papua New Guinea and West Africa, respectively, the power dynamics of triage, which place unequal social and political values on different lives. Triage is also shaped by expectations about what it means to be a good and compliant patient, client, citizen, farmer, or recipient of aid. We see this in Vincanne Adams's (2013) ethnography of the privatization of housing assistance and relief in post-Katrina New Orleans, wherein care, support, and recovery were made available to people along race and class lines—and selectively, at that. Donna Chollett (1996) applies Marilyn Gates's (1993) concept of "economic triage" to describe how Mexican sugarcane producers deemed more productive and efficient by processors were granted greater access to the market.

The fact that these methods of assessing, sorting, treating, and gatekeeping people emerged in the contexts of warfare, colonialism, global expansion of free trade and industrializing agriculture, deadly unnatural disasters, and epochs of economic austerity is eerily similar to actions in the era of the Trump administration and now amidst the coronavirus pandemic. One never knows what chaos lies ahead, but it has become terrifyingly predictable how such changes will affect the lives of those who are ecosocially vulnerable.

In the examples given herein, triage decision makers and systems designers hold a lot of power. Yet triage is not solely a tool of social control. Its forms and functions are not absolute or inflexible (Solomon 2017, citing Biehl 2013 and Han 2012). Methods of triage, as suggested by Mazak (2017), can be mobilized as strategies of social support and solidarity. Samuel Weeks (2015) describes how construction and day workers in Cape Verde use a triage system to delegate tasks and distribute jobs to the coworkers most in need in times of economic crisis and precarity. Farmworkers sometimes do this too, helping older or pregnant workers make their quotas. Harris Solomon describes how, in an Indian emergency room ward, the kin and kinfolk of injured patients work to move care along: "In scenes of triage, care is fractured and distributed. Kin, police and even perfect strangers attach themselves to a case and nudge it in both expected and unexpected directions" (2017, 361). This can, both subtly and significantly, challenge the authority of physicians and health care workers doing the initial sorting, potentially leading to better outcomes for patients and their families (Solomon 2017; see also Abramowitz et al. 2015). We can see this not-so-subtle nudging in the grassroots and activist responses to Hurricane Sandy, too, which contrasted with the "top-down, elite aid . . . characterized by finite programs with deadlines" (Superstorm Research Lab 2013, 3). For ordinary New York City area residents, the hurricane, and climate-change-induced problems more broadly, layered with other preexisting vulnerabilities—housing and food insecurity, unemployment and debt—long after the storm had subsided and thus required sustained social change work to address poverty and climate change injustices (Superstorm Research Lab 2013, 3, 16).

REPOLITICIZING TRIAGE IN ACTIVIST ANTHROPOLOGY

I have come to think of the different (and often invisible) nudging and accompanying—and the disruptive, pushing, and "participatory (re)action" (Maskens and Blanes 2013, 266) roles we take on as activist anthropologists in both our personal and professional lives—as a form of ecosocial triage. By appropriating and reconfiguring the temporal, spatial, and ethical designs of triage, engaged scholars, activists, and everyday citizens strategically suture together research, teaching, service, policies, material supports, health care, kinships, and other relationships and community memberships in home and field communities in new, unconventional, and critically imaginative ways. These invisible, emotional, and humble labors are ghostly. They are part of anthropology's activist history, which is too often erased from institutional and disciplinary memory. They are also hidden, because emotional labor and expressions are not valued or appreciated by all, including employing institutions and especially those with much to gain through the maintenance of the status quo. Our ghostly activist labors include engaged listening; applying medical anthropological insights and frameworks to collective efforts to develop more equitable and humane systems of care, food, and farming; law and knowledge sharing; rapid response networks. These are germane to all kinds of situations, including those I have described throughout *The Devil's Fruit* and others that readers have perhaps been involved with.

The challenge of triage is that amid one disaster there can be many, and even as a crisis calms or subsides, another erupts. Slow violence, where harms evolve gradually and at times invisibly, is multilayered and can be difficult to link back to a specific cause (Davies 2019; Nixon 2011). Here the labors of triage extend indefinitely, not unlike Krieger's fractal model of intertwining branches that represent the different dimensions, interconnections, and patterns of ecosocial health and disease (1994, 897). When dealing with the sometimes imperceptible and intergenerational harms of pesticide exposure, treatment may be delayed, denied, or deemed unnecessary—until it is too late. This is why Max Liboiron, Manuel Tironi, and Nerea Calvillo call for "forms of slow, intimate activism based in ethics rather than achievement." They add, "Slow activism does not have to be immediately affective or effective, premised on an anticipated result. It can just be good" (2018, 331, 341). The ethics of engaged ethnography are well suited to these tasks and could be worked into forms of ecosocial triage that attend to the uncertainties and imperceptibilities of toxic layering and invisible harm. We can do this by building and sustaining our relationships and maintaining our commitment and attention to the issues that affect the lives of our *compañeros/as*.

While triage helps instigate an immediate call to action, it can also make imagining—let alone building—systemic change more challenging. These responsibilities and labors exist amid the silencing chill of fascism, the uncertainties of "living on a damaged planet" (Gan et al. 2017, G1) and in the throes of a deadly global pandemic. Being bombarded with new information while just trying to sustain oneself and one's family or community emotionally, socially, and physically on a daily basis is exhausting. Maintaining a professional and personal balance between social change work and social service work (Kivel 2017) is never so straightforward. Community activists whom I've worked with throughout California lament their frustration at not being able to build or sustain momentum around a number of issues, from police violence and ICE raids and detentions to fighting one pesticide after the next. We are always engaged in the labor of what one activist called "putting out fires." (That California is also literally always on fire, and is running out of water, doesn't help.)

Activist anthropologist Faye Harrison (2010) speaks of "the pact" that anthropologists have with the communities we do research with, but this also applies to a pact we must have with ourselves as engaged anthropologists to support one another and our work. After reading about what the pact meant in *The Devil's Fruit*, what could it look like for you, the reader? How can you apply myth busting, or following and defamiliarizing objects, ideas, policies, or everyday assumptions in your community? What kinds of commitments—emotional, professional, and otherwise—are necessary for identifying and reacting to toxic layering and invisible harm where you live and work? What would accompaniment look and feel like in a difficult situation that you've experienced or witnessed? How can we relate and align ourselves more responsively to people we are distant from but also intimately connected to through the act of eating? How could a more ecosocial way of thinking and reacting, through research and activism, address problems in your lives and communities?

Admittedly, it is not always easy to engender this holism, either professionally or personally. We may get caught in intense political and disciplinary debates and conflicts. We may feel emotionally overwhelmed or personally or professionally vulnerable. We may not always have the privilege of taking a job that aligns with our personal ethics or values or to be vocal and active without putting ourselves at risk. As ethnographers, activists, or activist anthropologists we may or may not share histories, geographies, language, racial and ethnic identities, or occupations with the people we do research with. Yet when we put care and the "primacy of the ethical" (Scheper-Hughes 1990, 1995) first, we can make contributions to communities at individual and relational levels, as well as systemically. But we have to remember where we come from, remember the work of our anthropological activist ancestors, and remember to become haunting ourselves (Gan 2017). This requires remaining vigilant, and staying engaged in explicit and less overtly public ways in order to build alternative, healthy, and ecologically just futures with farmworkers and other *compañeros/as*.

ACKNOWLEDGMENTS

So many people have supported, guided, cheered, pushed, and consoled me during my research and writing work. I have done my best to remember and honor them here: as grains of sand and sustained flows, and amid the work of *The Devil's Fruit* that carries onward.

I thank Lenore Manderson, the series editor for Rutgers University Press's Medical Anthropology: Health, Inequality, and Social Justice, for her great faith in me and enthusiasm about this project. Throughout the writing and publishing process she provided helpful, caring, and constructive advice, and a gentle patience that I desperately needed while struggling to balance my responsibilities at a teaching-intensive public university along with personal grief and the weight of the world and this book on my shoulders. Rutgers anthropology editor Kimberly Guinta also believed in this project from its inception and gave me pushes and practical advice that I needed as a novice book author. Insightful feedback from anonymous reviewers made this manuscript even stronger. Thank you to Caroline Piroleau at the University of California–San Francisco's Memory and Aging Center for her help preparing the maps and images. Indigenous farmworker, activist, and photographer Amadeo Sumano provided many of the powerful photos featured in (and on the cover) of *The Devil's Fruit*. He is motived by a desire to share with the world what farm work is like from the point of view of *los campesinos/as* (farmworkers). Thorough production editing by Sherry Gerstein, and indexing by Kathleen Paparchontis have made this book clearer and easier to read and navigate.

I feel very fortunate to have a wide web of nurturing, supportive, and unabashedly radical intellectual kin and friends. They include Adrienne Pine and Brett Williams, whose courage as public scholars, anthropologists, and organizers continues to inspire me as an activist academic. The late feminist archeologist Joan Gero, Lesley Gill, Kimberly Grimes, Marco Hernandez, Dolores Koenig, Bryan McNeil, Mieko Nishida, Sabiyha Prince, Esperanza Roncero, Dolores Rowe, Eric Schott, Debarati Sen, Sue Taylor, David Vine, Rachel Watkins, and Mike Woost deserve praise for their roles in my long educational journey from grade school to PhD. I am also grateful to the Spanish teachers and host families affiliated with the Centro Internacional de Lenguas, Arte, y Cultura, or CILAC Freire, in Cuernavaca, Morelos, México.

I thank John Borrego, Dave Runsten, Adam Sanders, Sandy Brown, Sean Sweezy, Don Villarejo, Pat Zavella, and especially Ann López and the Center for Farmworker Families for orienting me to fieldwork with farmworkers on California's Central Coast. Anthropology colleagues Vincanne Adams, Seth Holmes,

Sarah Horton, Barbara Rose Johnston, and Angela Stuesse also provided moral, brainstorming, and collegial support in person and virtually. Lois Stanford gave me the tip to follow the strawberries during an elevator ride at my first American Anthropological Association Annual Meeting in 2008.

When I was a postdoctoral researcher at the Social Science Environmental Justice Research Institute at Northeastern University, Len Albright, Phil Brown, Max Liboiron, Jacob Matz, Miguel Montolva, Monica Ramirez-Andreotta, Mia Renauld, Lauren Richter, Laura Senier, Sara Shostak, Elisabeth Wilder, and Sara Wylie helped by reading very early article and chapter drafts. I am lucky to have learned from a group so dedicated to interdisciplinary, community-based activist scholarship.

The Pájaro and Salinas Valleys are home to many incredible and brilliant people. Their presence, generosity, and excitement about my work is laced throughout these pages. My thanks go to Luis Anguiano; Consuelo Alba; Mavel Armijo; Kathy and Duncan Blue; Cindy Fabry; Mary Flodin; Timothy Flynn and his late wife, Terry Ellis; Lowell Hurst; Breeze Medina and family, the late activist Cecile Mills, Jovita Molina, Path Star, and Pam Sexton and family. Special thanks also go to Kevin Cameron, Jacob Martinez, Juan Morales, and others at the Digital Nest, as well as my comrades in the Watsonville Autonomous Brown Berets and Watsonville Bike Shack Cooperative: Tomas Alejo, Manny Ballestranos, Lizette Bedolla-Cruz, Yovanna Bravo, Jorge Flores, Nayeli Gil, Sandino Gomez, Lorenzo Holquin, Jenn Laskin, Sal Lua, Ramiro Medrano, Yesenia Molina, Rocio Natividad, Frances Salgado-Chavez, and Sep Susunaga. I am also very appreciative of the teachers and administrators in the Pájaro Valley Unified School District Migrant Education District XI—namely, Veronica Fernandez, Rosa Hernandez, and Faris Sabbah. My thanks also go to the members of the Pájaro Valley Federation of Teachers—especially Sarah Henne, Jenn Laskin, and Francisco Rodriguez. Research assistants Yovanna Bravo and Silvia Perez helped me transcribe thousands of pages of English- and Spanish-language interviews and provided analysis and insights as the daughters and granddaughters of farmworkers. Staff from Californians for Pesticide Reform, Pesticide Action Network North America, Pesticide Watch, and Safe Ag Safe Schools—namely, Tracey Brieger, Kathryn Gilje, Susan Kegley, Dana Perls, Margaret Reeves, Paul Towers, and Mark Weller— supported Pájaro and Salinas Valley community efforts during the campaign against methyl iodide.

I am forever indebted to the many *campesinos/as* and *compañeros/as* who shared their knowledge and experiences with me and allowed me into their homes and lives. Several lawyers, doctors, nurses, farmers, growers, nonprofit directors and staff members, current and retired teachers, migrant educators, labor leaders, social workers, community organizers and activists, and social service providers also shared their perspectives with me and gave me opportunities to observe and engage. All of the people featured in this book have been given pseudonyms.

I thank my Fresno State University colleagues Jenny Banh, John Beynon, Kris Clarke, Hank Delcore, Anabella España-Navarro, Kathryn Forbes, Ken Hansen, T. Hassan Johnson, Hillary Jones, Chelsey Juarez, Lorretta Kensinger, Larissa Mercado-López, John Pryor, Jennifer Randles, DeAnna Reese, Aimee Rickman, Meta Schettler, Davorn Sisavath, Chris Sullivan, Everett Vieira, Julie Watson, and deans Michelle DenBeste and Luz Gonzalez. My many wonderful and brilliant Fresno State students are too numerous to name individually, but their energies and insights inspire me to carry on. Bonnie Bade at California State University San Marcos has been an anthropological *madrina* to me within the CSU system.

Good friends and peers from many different eras of my life, scattered all over the globe, have been consistent cheerleaders—especially Sara Artes, Abby Conrad, Michelle Glowa, Craig Hughes, Julie Koppel-Maldonado, Kira McKernan, Sarah Otto, Tracy Perkins, Jennie Simpson, Micah Trapp, Alice Brooke Wilson and Tom Philpott, and Morgan Windram-Geddes. In the Central Valley, I am lucky to have supportive relationships with Carissa Garcia and family, the Halloran-Espinoza family, Marci and Kiel Lopez-Schmitt, Liz Torres, Sarah Ramirez and David Terrell, Adrianna Alejo Sorondo, Gloria Hernanrez, Rey León, Irene Parra Serrano and her dear late sister Irma Serrano, Devoya Mayo, Christina Alejo, Ariana Martinez Lott, Chucho Mendoza, and sisters Grisanti and Yenedit Valencia and family. Thanks are due also to the folks at the Fresno Freedom School—especially the late Maria Else, Floyd Harris, Aline Reed, Ivanka Sanders-Hunt, and all the kids.

I am extremely fortunate to be an Atlantic Fellow for Equity in Brain Health with the Global Brain Health Institute at University of California–San Francisco for the 2019–2020 year. I appreciate the time and space this opportunity has afforded me to wrap up this book and to design a follow-up project on farmworkers' lived and embodied experiences of aging in toxic places in California's Central Valley. Special thanks go to Gloria Aguirre, Alissa Bernstein, Dominic Campell, Bruce Miller, Victor Valcour, Kristine Yaffe, and the thirty-two phenomenal fellows in my cohort for their support—especially Karin Diamond, Marcela Mar, and Jennie Gubner.

My dad, Ronald L. Saxton, passed away as I came close to finishing this manuscript. His influence, encouragement, wisdom, and community legacies are thoroughly ingrained in me and the pages of this book. I miss him, and I wish he could read this. My mom, Vera E. Kaminski Saxton, cultivated my creative streaks, the courage question, and an almost indefatigable work ethic. My siblings, Elena and Ron E. Saxton, provided moral support, rides to and from the airport, and encouragement.

Portions of this book were previously published in "Ethnographic Movement Methods: Anthropology Takes on the Pesticide Industry," *Journal of Political Ecology* 22, no. 1: 368–388; "Layered Disparities, Layered Vulnerabilities: Farmworker Health and Agricultural Corporate Power on and off the Farm" (PhD diss., American University, 2013); and "Strawberry Fields as Extreme Environments: The Ecobiopolitics of Farmworker Health," *Medical Anthropology* 34, no. 2: 166–183. My research has been supported by funding from the American University College of Arts and Sciences Doctoral Dissertation Research Award (2010–2011), the American University Provost's Dissertation Writing Award (2013), a National Science Foundation Doctoral Dissertation Research Improvement Grant (Award no. 1059537), and the Wenner-Gren Foundation (Grant no. 8269). Many semesters of research release time from the Fresno State College of Social Sciences also supported me while completing this book.

All errors, limitations, and omissions are my own.

NOTES

INTRODUCTION

Epigraphs: Cartwright and Manderson 2011, 453; Salomón J. 2015, 86.

1. A shortcoming of this book is to speak of the legacy of activist anthropology in North America while neglecting the many stories of engaged, public, and activist anthropology in other parts of the world and in the areas of linguistic, archeological, and biophysical and forensic anthropology.

2. Whistleblowing and other forms of explicit or overt activism can come with personal and professional risks that may significantly impact one's ability to continue to do one's work in the future. More tacit or institutionally accepted forms of activist anthropology could include anthropologists working from within state, private, corporate, nongovernmental, or nonprofit institutions to mitigate environmental and other health hazards or sharing the results of their studies with activists without directly participating in movements. For further discussion, see Johnston 2001.

3. See, for example, Bartley et al. 2015; Besky 2013; DuPuis 2002; Freidberg 2010; Gray 2013; Guthman 2004; Lyon and Moberg 2010; Schlosser 2001; and Szasz 2007.

4. See, for example, the many entries on *Access Denied*, a blog by critical and activist medical anthropologists, about how care is denied in formal institutions of caring like hospitals and clinics. See *Access Denied* (blog), https://accessdeniedblog.wordpress.com/recent-posts-3/.

5. Other researchers have demonstrated how things like organic and fair trade standards and labeling schemes have too often been co-opted by the very corporate entities they were meant to resist (Brown and Getz 2008; Daria 2019; Gray 2013; Guthman 2004; Lyon and Moberg 2010; Szasz 2007). Organic and fair trade endeavors are important, but we must be mindful that diverse values and motivations condition what outcomes they have for producers and workers.

1 ENGAGED ANTHROPOLOGY WITH FARMWORKERS

1. In 1942, during World War II, the agricultural industry successfully lobbied for a formalized foreign farmworker program, arguing that the war would diminish their supplies of domestic workers. Between 1942 and 1964, millions of *braceros* (literally meaning farmhands, or people who work with their arms, *brazos*) from México received seasonal contracts to do farm labor jobs. Even after the war ended, the program continued. Then, as now, as labor historian Sarah Hines observes, "The primary purpose and result of the bracero program was to give growers more control over farm labor, immigrant and native alike. The program was immensely profitable for growers, as it enabled them to thwart union organizing efforts and drive down wages of all farm workers" (2006). The Bracero Program ended in 1964, but the H-2A program, established in 1952 as part of the Immigration and Nationality Act (Önel and Farnsworth 2016) remains and serves a similar function, especially amidst claims of an ongoing socially engineered labor shortage.

2. The current crisis at the Mexican-U.S. border involves mostly migrants from Central America who are fleeing political violence and severe food insecurity brought on by drought

and climate change, both of which are linked to U.S. foreign policy and economic exploitation in the region over the course of several decades.

3. Many Triqui people immigrated to the United States seeking political asylum and fleeing political violence in Oaxaca. For details, see Stephen 2017.

4. As Horton (2016b) observes, however, the synergistic relationships between poverty, stress, and the harsh and hot conditions of work are far from natural and can exacerbate and contribute to diseases like diabetes, hypertension and kidney disease.

5. César Chávez is also a contested figure in farmworker activism and history. See, for example, Bardacke 2011, M. Garcia 2014, and Neuburger 2013.

6. The California Institute for Rural Studies has made some progress in mobilizing the data it collected from its farmworker housing study to pressure city and county governments and some growers to build new subsidized or employer-provisioned units for farmworkers and their families. Other efforts at housing provided by grower-employers in the Salinas Valley have at times fallen short of community expectations. For example, the lettuce growing firm Tanimura and Antle built nice farmworker housing for temporary and migrant workers in Spreckels, California, outside of Salinas, but this is not available to longer-term resident farmworker families who struggle to find housing (Morehouse 2016).

7. Secure Communities became the Prioritized Enforcement Program under ICE from 2015 through 2017 and emphasized detaining and deporting people convicted of more severe and violent offenses. Under the Trump administration, the federal policy no longer "prioritizes" detention or deportation based upon the severity, repetition, or violent nature of the crime committed; instead, anyone who ends up in jail, irrespective of the reason or the person's individual history, can be subject to immediate detention and deportation through ICE (American Immigration Council 2018).

8. For another perspective on the racialized scapegoating of Chicano/a and Latinx communities for deforestation, see Kosek 2006.

9. Coincidentally, this individual was later arrested and convicted of fraud for misappropriating funds for a fake charitable organization that he had set up in order to improve the neighborhood and rid it of crime.

10. In other instances, labor contractors use "loaned identity documents" to "simultaneously obscure [undocumented workers'] presence from the state and federal governments while benefiting from the wage deductions associated with such prohibited workers' labor" (Horton 2016a, 11).

11. Jimenez's and Silva Ibarra's stories of being denied immediate medical attention by supervisors parallel what Aniceta (in chapter 4) and farmworkers exposed to pesticide drift (in chapter 3) and other injuries endure. Ultimately, several of Silva Ibarra's coworkers went on strike to protest their *compañero's* death and their unjust working conditions. They lost their jobs, and their H-2A visas were revoked; work stoppages are deemed a violation of their guest working contracts (Abell 2018). Yet, following procedures and making a formal complaint, even without a protest, could have resulted in a similar outcome. "If you don't like it, leave!" is a common refrain heard by farmworkers from their supervisors after bringing up concerns about things as basic as access to water or clean bathrooms.

12. A further complication in the case of permanent disabilities resulting from workplace injuries and illnesses is that different injured body parts—from hands to legs—are monetarily valued at variable rates in different states (Grabell 2015a, 2015b, 2015c, 2015d, 2015e, 2017; Grabell and Groeger 2015).

13. Long and hard-fought wage increases and rights for farmworkers secured by the UFW also faced rollbacks with the passage of the 1986 Immigration Reform and Control Act, which legalized millions of im/migrants who had been living and working in the United States. As

many newly documented farmworkers moved into nonfarm jobs, creating a labor shortage, growers responded by recruiting undocumented workers directly from Central America and México via farm labor contractors. Contractors exploit these workers' status and vulnerabilities through wage theft and labor abuses, sometimes not paying them at all. In some cases, threats of deportation and abuse, and in others, actual rape and sexual assault, are used to discipline farmworker crew members. We are seeing similar patterns with the renewed interest and utilization of the H-2A temporary worker program.

2 STRAWBERRIES

1. This parallels the California Department of Health Services' 5 a Day—For Better Health! campaign, established in 1988. The five-a-day (or sometimes even nine-a-day) mantra to increase fruit and vegetable consumption, and campaigns that encourage three servings a day of dairy, persist in advertisements featuring tight-bodied and jovial celebrities, seemingly sound medical advice, and public health dietary dictates, despite scant and mixed evidence that any one diet or food makes sense for everyone. They are not always supported by health data; instead they are heavily funded by agribusiness and food industries. The California Strawberry Commission, for example, has contracted medical doctors and scientists to tout the many health benefits of strawberries on its website, YouTube channel, and social media sites. For in-depth critiques of this kind of industry-sponsored and often state-endorsed pseudoscientific nutritionism, see Nestle 2013, 2018; see also Guthman 2011; and Wiley 2013.
2. Since the passage of the Immigration Reform and Control Act under President Ronald Reagan in 1986, political efforts to secure rights and/or amnesty for undocumented im/migrants in the United States have largely failed, even though the agricultural industry is highly dependent on their labor.
3. For a more thorough review of the history and evolution of the California strawberry and the strawberry industry, see Guthman 2019.
4. To date, no genetically modified varieties of strawberries, in which genetic materials from different species are crossed, have been grown commercially, though there has been some research and development, and field trials, with genetically modified strawberries (Husaini and Abdin 2008; James et al. 1993; Morgan et al. 2002).
5. Guthman (2019, 62-70) also describes how the differences between public and private domain in plant breeding in California are not always so clear cut. Public land grant universities like the University of California often partner with private companies to fund plant science and breeding research, and universities are increasingly behaving like private companies following the withdrawal of state and federal support for research in the public interest.
6. For example, the BugVac, created by Driscoll's in the 1980s, is one such nontoxic pest-control method. The device sucks up strawberry eating insects with a series of giant vacuums attached to a tractor that passes over the rows (The Packer 2015). Experiments are also ongoing using steam, heat, soil amendments, non-soil growing mediums, and predator microorganisms to discourage the flourishing of harmful pests (Guthman 2019; UC IPM 2018).
7. Consumers who purchase organic strawberries and think that by doing so they are opting out of contributing to these environmental and social harms are often surprised to learn that even organic strawberry transplants are nurtured in fumigant-treated soil. Many growers are skeptical of the commercial viability of unfumigated stock, and regulators worry about the potential of statewide pest outbreaks. The first and only fumigant-free nursery went out of business in California in 2000 (R. Gross 2011). The practice of treating soil used to grow starts, transplants, and saplings, organic or not, is widespread. Once the young plants arrive at individual farms, organic and conventional growers use different methods to maintain pest-

free crops, the main difference being that "certified organic" farmers can only use methods and substances that are approved by the U.S. Department of Agriculture's National Organic Program; certification allows them to market their products as organic. Some ecologically conscious farmers, however, resent the codification, commodification, and state regulation of agricultural knowledge; they still use organic methods, but cannot or will not sell their products under an "organic" label (Guthman 2004).

8. Partial or full mechanization of harvests has been standard practice in canning tomato and lettuce fields for several decades (Friedland 1984, 1994; Neuburger 2013), and is also common practice in California for melons (Horton 2016b), wine grapes, olives, and tree nuts.

9. For a discussion of how farmwork jobs in the United States of different degrees of physical intensity and different pay rates are assigned based on assumptions about different racial and ethnic groups and the rights that are respected when workers have U.S. citizenship or legal status, see Cartwright 2011; and Holmes 2007, 2013.

3 PESTICIDES AND FARMWORKER HEALTH

1. For an analysis of the various social and structural factors that prevent injured and ill workers from seeking and receiving care, see Snipes, Cooper, and Shipp 2017.

2. The updated U.S. EPA's WPS also requires extensive safety training for pesticide applicators and farmworkers, including mandatory notifications of recent applications and the use of personal protective equipment in certain circumstances. The year 2017 marked the first time in over four decades that the WPS was updated, and while it was significantly improved, it still does not substantially address farmworkers' concerns or reflect their lived experiences and realities (Bohme 2015a).

3. Emerging research suggests that farmworkers exposed to neurotoxic pesticides may experience a weakened sense of smell and that this can be an early indicator of future neurodegenerative diseases; see, for example, Quandt et al. 2017.

4. These include chemicals like chloropicrin, metam sodium, 1-3-D, and dazomet.

5. The U.S. Government Accountability Office notes significant limitations in the U.S. Food and Drug Administration's and the USDA's pesticide residue monitoring methodologies (L. Gross 2019; U.S. GAO 2014), suggesting that consumer exposure may also be grossly underestimated.

6. For a detailed and interactive map of pesticide use concentration, see the CalEnviroScreen 3.0 mapping tool (OEHHA, n.d.-b).

7. Other classes of pesticides include organochlorines, like DDT, which are largely no longer used in the United States but still linger in the environment; organofluorides; carbamates; and pyrethroids. In lay language, most people describe pesticides by emphasizing the pests or diseases they target, such as herbicides (to kill weeds), insecticides, fungicides, rodenticides, and nematicides, to name a few.

8. California is also the first state to institute its own research and regulatory initiative, Biomonitoring California. One goal of the program is to "help assess the effectiveness of public health efforts and regulatory programs to decrease exposures to specific chemicals," including several classes of pesticides (Biomonitoring California, n.d.). Measuring evidence of pesticide exposure in the body involves testing for the metabolite or breakdown products that appear in measurable amounts in certain bodily fluids. This raises the question of perceptibility again, since not all of the indicators or biomarkers of exposure are known or thoroughly understood (see Arcury et al. 2006). New, more accurate, precise, and thorough pesticide biomonitoring tests are needed.

9. For a comprehensive overview, see the Endocrine Society's "Second Scientific Statement on Endocrine-Disrupting Chemicals" (Gore et al. 2015).

4 ACCOMPANYING FARMWORKERS

1. Both Lynd (2013) and Farmer (2003, 2004, 2011) borrow heavily from Latin American liberation theologists, who have preached and practiced for the poor in their home communities and countries. See also Besteman 2015; Pine 2013; Sanjek 2015; and Williams 2015.
2. Some have suggested that smaller-scale, diversified organic farms may provide some occupational safety benefits to farmworkers if they are allowed to change tasks throughout the day (see, for example, Strolich et al. 2008, 18–19). In general, labor practices on organic, local, or small farms are not necessarily safer or more ethical than those found on large commercial farms (Gray 2012; Guthman 2004; Holmes 2013).
3. For a rich description of the syndemic health consequences of im/migration and sexual trauma and violence, see Mendenhall 2012.
4. For more details about the challenges faced by Mexican farmworkers turned farmers, as well as the vital contributions they are making to sustainable agriculture in California, see Minkoff-Zern 2019.
5. Empty Bowl fundraisers are hosted all over the United States via Feeding America, a major nonprofit organization that works with the state, growers, and grocers to consolidate the flow of donated food into food banks (Fisher 2018).
6. Sarah Horton (2016b) describes how she assumed the combined role of anthropologist and social worker in her field site in Mendota, California, and became a *raitera* (ride) for farmworkers who lived over forty miles away from the nearest hospital and other social services.

5 ECOSOCIAL SOLIDARITIES

1. In 2014 Permira sold its stake in Arysta LifeScience to the chemical firm Platform Specialty Products (Bray 2014). Arysta was sold again in 2018 to UPL Limited for $4.2 billion, making it the fifth largest pesticide company globally (Rusnak 2018).
2. This is similar to what Scheper-Hughes (1992) observed when doctors in Brazil medicalized the social suffering and emotional distress expressed by parents of starving children in shantytowns, offering pills to temporarily calm their nerves and quell hunger instead of plans to improve area food security.
3. Reentry intervals are set by the CA DPR and are intended to prevent workers and passersby from entering a field where pesticides have recently been applied. But farmworkers I interviewed indicated that reentry intervals were not always enforced, and the policies did little to protect rural residents, students, teachers, or school staff who lived or worked adjacent to farm fields.
4. Not all doctors' offices provide such comprehensive health care with respect to patient education on pesticide exposure risks. The American College of Obstetricians and Gynecologists, Physicians for Social Responsibility, and Planned Parenthood, as well as unionized nurses, were among the few health organizations and health care providers to join students and others in the struggle against methyl iodide. Local nonprofit clinics and some private practices specializing in occupational medicine, often financed by grants and donations from or patronized by agribusinesses, have remained overwhelmingly silent on the issue beyond

instructing workers to take precautions at work (by washing hands and not eating fruit in the fields) and at home (by laundering work clothes separately from other clothes and keeping a clean house).

5. Yield differences between organic and conventional production in strawberries and other crops is debated and conflicted in the literature; it varies by region, growing conditions, and farming techniques; see Guthman 2019; and Seufert, Ramankutty, and Foley 2012.

6. For mapping data about school and residential proximity to agricultural fields in Monterey and Tulare Counties, California, see S. J. Steinberg and S. L. Steinberg 2008; and S. L. Steinberg and S. J. Steinberg 2008.

7. A significant exception to this was the $289 million settlement that Monsanto, the manufacturer of the common agricultural and household herbicide Roundup (glyphosate), had to pay to former Benicia, California, school groundskeeper Dewayne Johnson (Levin 2018).

REFERENCES

Abell, Evan. 2018. "Temporary Farmworker Talks about Working Conditions on Sumas Berry Farm, Why He Came to U.S." *Bellingham (WA) Herald*, February 6, 2018. https://www.bellinghamherald.com/news/local/article167132492.html.

Abramowitz, Sharon Alane, Kristen E. McLean, Sarah Lindley McKune, Kevin Louis Bardosh, Mosoka Fallah, Josephine Monger, Kodjo Tehoungue, and Patricia A. Omidian. 2015. "Community-Centered Responses to Ebola in Urban Libera: The View from Below." *PLoS Neglected Tropical Diseases* 9, no. 4.

Ackerknecht, Erwin H. 1953. *Rudolf Virchow: Doctor, Statesmen, Anthropologist*. Madison: University of Wisconsin Press.

ACLU Northern California. n.d. "TRUST Act (AB 4)." Accessed September 7, 2020. https://www.aclunc.org/our-work/legislation/trust-act-ab-4.

Adams, Vincanne. 2013. *Markets of Sorrow, Labors of Faith: New Orleans in the Wake of Katrina*. Durham, NC: Duke University Press.

Adeyinka, Adebayo, and Lousidon Pierre. 2019. "Organophosphates." In *StatPearls*, online ed. Treasure Island, FL: StatPearls. https://www.ncbi.nlm.nih.gov/books/NBK499860/.

Aggarwal, Neil. 2008. "Editorial: Farmer Suicides in India: The Role of Psychiatry and Anthropology." *International Journal of Social Psychiatry* 54, no. 4: 291–292.

Ahn, Christine. 2017. "Democratizing American Philanthropy." In *The Revolution Will Not Be Funded: Beyond the Non-Profit Industrial Complex*, edited by INCITE!, 63–78. Durham, NC: Duke University Press.

Alavanja, Michael. C., Mary H. Ward and Peggy Reynolds. 2007. "Carcinogenicity of Agricultural Pesticides in Adults and Children." *Journal of Agromedicine* 12, no. 1: 39–56.

Alewu, B., and Chidi Nosiri. 2011. "Pesticides and Human Health." In *Pesticides in the Modern World: Effects of Pesticides Exposure*, edited by Margarita Stoytcheva, 205–231. London: InTech. http://www.intechopen.com/books/pesticides-in-the-modern-world-effects-of-pesticides-exposure/pesticide-and-human-health.

Alexander, Dominik. 2013. "Epidemiological Critique of Gemmill et al. 2013, Residential Proximity to Methyl Bromide Use and Birth Outcomes in Agricultural Populations in California." White paper. Boulder, CO: Exponent Inc. Health Sciences, 2013.

Ali, Kamran Asdar. 2010. "Voicing Difference: Gender and Civic Engagement among Karachi's Poor." *Current Anthropology* 51, no. S2: S313–S320.

American Immigration Council. 2018. "The End of Immigration Enforcement Priorities under the Trump Administration." Fact sheet, March 7, 2018. https://www.americanimmigrationcouncil.org/research/immigration-enforcement-priorities-under-trump-administration.

Andersen, Barbara. 2016. "Temporal Circuits and Social Triage in a Papua New Guinean Clinic." *Critique of Anthropology* 36, no. 1: 13–26.

Anger, Scott, and Sasha Khokha. 2013. *Hunger in the Valley of Plenty: Hungry in Raisin City*. San Francisco: KQED / Center for Investigative Reporting. YouTube, https://www.youtube.com/watch?v=_ZTxYVdMfLs.

APHA (American Public Health Association). 2010. "Requiring Clinical Diagnostic Tools and Biomonitoring of Exposures to Pesticides." Web page, November 9, 2010. https://www.apha.org/policies-and-advocacy/public-health-policy-statements/policy-database

/2014/07/21/09/17/requiring-clinical-diagnostic-tools-and-biomonitoring-of-exposures -to-pesticides.

Appadurai, Arjun. 1986. "Introduction: Commodities and the Politics of Value." In *The Social Life of Things: Commodities in Cultural Perspective*, edited by Arjun Appadurai, 3–63. Cambridge: Cambridge University Press, 1986.

Arax, Mark. 2019. *The Dreamt Land: Chasing Water and Dust Across California*. New York: Alfred A. Knopf.

———. 2005. *The King of California: J. G. Boswell and the Making of a Secret American Empire*. New York: Public Affairs Books.

Arcury, Thomas A., Joseph G. Grzywacz, Haiying Chen, Quirina. M. Vallejos, Leonard Galván, Laura Whalley, Scott Isom, Dana B. Barr, and Sara A. Quandt. 2009. "Variation across the Agricultural Season in Organophosphorus Pesticide Urinary Metabolite Levels for Latino Farmworkers in Eastern North Carolina: Project Design and Descriptive Results." *American Journal of Industrial Medicine* 52, no. 7: 539–50. https://doi.org/10.1002 /ajim.20703.

Arcury, Thomas A., Ilene J. Jacobs, and Virginia Ruiz. 2015. "Farmworker Housing Quality and Health." *New Solutions: A Journal of Environmental and Occupational Health Policy* 25, no. 3: 256–262.

Arcury, Thomas A., Ha T. Nguyen, Phillip Summers, Jennifer W. Talton, Lourdes Carrillo Holbrook, Francis O. Walker, Haiying Chen, et al. 2014. "Lifetime and Current Pesticide Exposure among Latino Farmworkers in Comparison to Other Latino Immigrants." *American Journal of Industrial Medicine* 57, no. 7: 776–787.

Arcury, Thomas A., and Sara A. Quandt. 2003. "Pesticides at Work and at Home: Exposure of Migrant Farmworkers." *Lancet* 362, no. 9400: 20–21.

Arcury, Thomas A., Sara A. Quandt, Dana B. Barr, Jane A. Hoppin, Linda McCauley, Joseph G. Grzywacz, and Mark G. Robson. 2006. "Farmworker Exposure to Pesticides: Methodologic Issues for the Collection of Comparable Data." *Environmental Health Perspectives* 114, no. 6: 923–928.

Arcury, Thomas, Sara A. Quandt, and Gregory B. Russell. 2002. "Pesticide Safety among Farmworkers: Perceived Risk and Perceived Control as Factors Reflecting Environmental Justice." *Environmental Health Perspectives* 110, no. 2: 233–240.

Associated Press. 2018. "State Agency: Sumas Farm Not at Fault in Farmworkers' Death." *Seattle Times*, February 1, 2018. https://www.seattletimes.com/seattle-news/state-agency-sumas -farm-not-at-fault-in-farmworkers-death/.

ATSDR (Agency for Toxic Substances and Disease Registry). 2020. *Toxicological Profile for Bromomethane*. Atlanta: ATSDR. https://www.atsdr.cdc.gov/ToxProfiles/tp27.pdf.

Atwood, Donald, and Claire Paisley-Jones. 2017. *Pesticides Industry Sales and Usage: 2008–2012 Market Estimates*. Washington, DC: U.S. Environmental Protection Agency, 2017. https:// www.epa.gov/sites/production/files/2017-01/documents/pesticides-industry-sales -usage-2016_0.pdf.

Aviv, Rachel. 2014. "A Valuable Reputation." *New Yorker*, February 10, 2014. https://www .newyorker.com/magazine/2014/02/10/a-valuable-reputation.

Backe, Emma Louise. 2017. "Engagements with Ethnographic Care." *Anthropology News* 58, no. 1: e81–e86.

Bacon, David. 2013. "Yesterday's Internment Camp—Today's Labor Camp." Truthout, September 16, 2013. https://truthout.org/articles/yesterdays-internment-camp-todays-labor -camp/.

Baer, Hans A., Merrill Singer, and Ida Susser. 2003. *Medical Anthropology and the World System*. 2nd ed. Westport, CT: Praeger.

Baer, Roberta D., and Dennis Penzell. 1993. "Susto and Pesticide Poisoning among Florida Farmworkers." *Culture, Medicine, and Psychiatry* 17: 321–327.

Bail, Kari M., Jennifer Foster, Saifiya George Dalmida, Ursula Kelly, Maeve Howett, Erin P. Perranti, and Judith Wold. 2012. "The Impact of Invisibility on the Health of Migrant Farmworkers in the Southeastern United States: A Case Study from Georgia." *Nursing Research and Practice* 2012 (1): epub 760418.

Bank Muñoz, Carolina. 2008. *Transnational Tortillas: Race, Gender and Shop-Floor Politics in Mexico and the United States.* Ithaca, NY: Cornell University Press, 2008.

Bardacke, Frank. 2011. *Trampling Out the Vintage: Cesar Chavez and the Two Souls of the United Farm Workers.* New York: Verso.

———. 1994. *Good Liberals and Great Blue Herons: Land, Labor and Politics in the Pájaro Valley.* Santa Cruz, CA: Center for Political Ecology.

Barndt, Deborah. 2008. *Tangled Routes: Women, Work and Globalization on the Tomato Trail.* 2nd ed. Lanham, MD: Rowman and Littlefield.

Bartley, Tim, Sebastian Koos, Hiram Samel, Gustavo Setrini, and Nik Summers. 2015. *Looking behind the Label: Global Industries and the Conscientious Consumer.* Bloomington: Indiana University Press.

Bassil, Kate L., Cathy Vakil, Margaret Sanborn, Donald C. Cole, Judith S. Kaur, and Kathleen J. Kerr. 2007. "Cancer Health Effects of Pesticides: Systematic Review." *Canadian Family Physician* 53, no. 10: 1704–1711.

Bauer, Kathleen. 2019. "Rosalinda Guillen Is a Force for Farmworker Justice." Civil Eats, July 1, 2019. https://civileats.com/2019/07/01/rosalinda-guillen-is-a-force-for-farmworker-justice/.

Benson, Peter. 2011. *Tobacco Capitalism: Growers, Migrant Workers and the Changing Face of a Global Industry.* Princeton, NJ: Princeton University Press.

———. 2008a. "El Campo: Faciality and Structural Violence in Farm Labor Camps." *Cultural Anthropology* 23, no. 4: 589–629.

———. 2008b. "Good Clean Tobacco: Philip Morris, Biocapitalism and the Social Course of Stigma in North Carolina." *American Ethnologist* 35, no. 3: 357–379.

Benson, Peter, and Stuart Kirsch. 2010. "Capitalism and the Politics of Resignation." *Current Anthropology* 51, no. 4: 459–486.

Bergman, Åke, Jerrold J. Heindel, Tim Kasten, Karen A. Kidd, Susan Jobling, Maria Neira, R. Thomas Zoeller, et al. 2013. "The Impact of Endocrine Disruption: A Consensus Statement on the State of the Science." *Environmental Health Perspectives* 121, no. 4: A104–A106.

Berk, Marc L., Claudia L. Schur, Leo R. Chavez, and Martin Frankel. 2000. "Health Care Use among Undocumented Latino Immigrants." *Health Affairs* 19, no. 4: 51–64.

Berkey, Rebecca E. 2017. *Environmental Justice and Farm Labor.* Abingdon, UK: Routledge.

Bernstein, Hamutal, Dulce Gonzalez, Michael Karpman, and Stephen Zuckerman. 2019. *One in Seven Adults in Immigrant Families Reported Avoiding Public Benefit Programs in 2018.* Washington, DC: Urban Institute. https://www.urban.org/research/publication/one-seven-adults-immigrant-families-reported-avoiding-public-benefit-programs-2018.

Besky, Sarah. 2018. "Sickness." Society for Cultural Anthropology, July 26, 2018. https://culanth.org/fieldsights/sickness.

———. 2017. "Monoculture." Society for Cultural Anthropology, June 28, 2017. https://culanth.org/fieldsights/monoculture.

———. 2013. *The Darjeeling Distinction: Labor and Justice on Fair-Trade Tea Plantations in India.* Berkeley: University of California Press.

Besky, Sarah, and Alex Blanchette. 2018. "Introduction: The Naturalization of Work." Society for Cultural Anthropology, July 26, 2018. https://culanth.org/fieldsights/introduction-the-naturalization-of-work.

Besteman, Catherine. 2015. "On Ethnographic Love." In *Mutuality: Anthropology's Changing Terms of Engagement*, edited by Roger Sanjek, 259–284. Philadelphia: University of Pennsylvania Press.

Bhopal Medical Appeal. 2016. "Sambhavna Study Says Gas Hit Ten Times More Prone to Cancer." Bhopal Medical Appeal, September 5, 2016. http://www.bhopal.org/sambhavna-study-says-gas-hit-ten-times-more-prone-to-cancer/.

———. 2013. "New Abnormalities Seen in Third Generations of Bhopal Children." Bhopal Medical Appeal, March 4, 2013. http://www.bhopal.org/new-abnormalities-seen-in-third-generations-of-bhopal-children/.

———. n.d. "Bhopal's Second Poisoning." Bhopal Medical Appeal. Accessed September 7, 2020. https://www.bhopal.org/second-poisoning/bhopal-second-poisoning/.

Biehl, João. 2013. *Vita: Life in a Zone of Social Abandonment*. Berkeley: University of California Press.

Biomonitoring California. n.d. "Biomonitoring California." Home page, accessed September 7, 2020. https://biomonitoring.ca.gov/.

Biruk, Crystal. 2017. "Ethical Gifts? An Analysis of Soap-for-Data Transactions in Malawian Survey Research." *Medical Anthropology Quarterly* 31, no. 3: 365–384.

Bloom, John. 2011. "'The Farmers Didn't Particularly Care for Us': Oral Narrative and the Grass Roots Recovery of African American Migrant Farm Labor History in Central Pennsylvania." *Pennsylvania History: A Journal of Mid-Atlantic Studies* 78, no. 4: 323–354.

Bohme, Susanna Rankin. 2015a "EPA's Proposed Worker Protection Standard and the Burdens of the Past." *International Journal of Occupational and Environmental Health* 21, no. 2, 161–165.

———. 2015b. *Toxic Injustice: A Transnational History of Exposure and Struggle*. Berkeley: University of California Press.

Borrego, John, and Patricia Zavella. 1999. "Policy Implications of the Restructuring of Frozen Food Production in North American and Its Impact on Watsonville, California." Working Paper No. 28, Chicano/Latino Research Center, University of California–Santa Cruz.

Bottemiller Evich, Helena, Jenny Hopkinson, and Eric Wolff. 2017. "Dems' 'Blue Card' Bill Would Protect Ag Workers." *Politico*, May 4, 2017. https://www.politico.com/tipsheets/morning-agriculture/2017/05/dems-blue-card-bill-would-protect-ag-workers-220131.

Bourgois, Philippe. 1990. "Confronting Anthropological Ethics: Ethnographic Lessons from Central America." *International Journal of Peace Research* 27, no. 1: 43–54.

———. 2009. "Recognizing Invisible Violence: A Thirty-Year Ethnographic Retrospective." In *Global Health in Times of Violence*, edited by Barbara Rylko-Bauer, Linda Whiteford, and Paul Farmer, 18–40. Santa Fe, NM: School of Advanced Research.

———. 2006. "Foreword: Anthropology in the Global State of Emergency." In *Engaged Observer: Anthropology, Advocacy, and Activism*, edited by Victoria Sanford and Asale Angel-Ajani, ix–xii. New Brunswick, NJ: Rutgers University Press, 2006.

Bowen, Monica. 2011. "Strawberries as an 'Earthly Delight.'" *Alberti's Window* (blog), July 9, 2011. http://albertis-window.com/2011/07/strawberries-as-an-earthly-delight/.

Boyer, Dominic. 2015. "Reflexivity Reloaded: From Anthropology of Intellectuals to Critique of the Method to Studying Sideways." In *Anthropology Now and Next: Essays in Honor of Ulf Hannerz*, edited by Thomas Hylland Erickson, Christina Garsten and Shalini Randeria, 91–110. New York: Berghahn Books.

Bradman, Asa, Alicia L. Salvatore, Mark Boeniger, Rosemary Castorina, John Snyder, Dana B. Barr, Nicholas P. Jewell, Geri Kavanaugh-Baird, Cynthia Striley, and Brenda Eskenazi. 2009. "Community-Based Intervention to Reduce Pesticide Exposure to Farmworkers and Potential Take-Home Exposure to Their Families." *Journal of Exposure Science and Environmental Epidemiology* 19, no. 1: 79–89.

Bradman, Asa, Donald Whitaker, Lesliam Quirós, Rosemary Castorina, Birgit Claus Henn, Marcia Nishioka, Jeffrey Morgan, Dana B. Barr, Martha Harnly, Judith A. Brisbin, Linda S. Sheldon, Thomas E. McKone, and Brenda Eskenazi. 2007. "Pesticides and Their Metabolites in the Homes and Urine of Farmworker Children Living in the Salinas Valley, CA." *Journal of Exposure Science and Environmental Epidemiology* 17, no. 4: 331–349.

Bray, Chad. 2014. "Platform Specialty Products to Pay $3.5 Billion for Arysta LifeScience." *New York Times*, October 20, 2014. https://dealbook.nytimes.com/2014/10/20/platform -specialty-products-to-pay-3-5-billion-for-arysta-an-insecticide-maker/.

Brehm, Emily, and Jodi A. Flaws. 2019. "Transgenerational Effects of Endocrine-Disrupting Chemicals on Male and Female Reproduction." *Endocrinology* 160, no. 6: 1421–1435.

Brown, Kate, and Olha Martynyuk. 2016. "The Harvests of Chernobyl." *Aeon*, November 29, 2016. https://aeon.co/essays/ukraine-s-berry-pickers-are-reaping-a-radioactive-bounty.

Brown, Phil. 2007. *Toxic Exposures: Contested Illnesses and the Environmental Health Movement.* New York: Columbia University Press.

Brown, Sandy, and Christy Getz. 2011. "Farmworker Food Insecurity and the Production of Hunger in California." In *Cultivating Food Justice: Race, Class and Sustainability*, edited by Alison Hope Alkon and Julian Agyeman, 121–146. Cambridge, MA: MIT Press.

———. 2008. "Privatizing Farm Worker Justice: Regulating Labor through Voluntary Certification and Labeling." *Geoforum* 39, no. 3: 1184–1196.

Brown, Terry P., Paul C. Rumsby, Alexander C. Capleton, Lesley Rushton, and Leonard S. Levy. 2006. "Pesticides and Parkinson's Disease—Is There a Link?" *Environmental Health Perspectives* 114, no. 2: 156–164. https://www.ncbi.nlm.nih.gov/pmc/articles /PMC1367825/.

Bruggmann, Mary Wallace. 1966. "The Strawberry in Religious Paintings of the 1400s." In *The Strawberry: History, Breeding and Physiology*, edited by George M. Darrow, 11–14. New York: Holt, Rinehart and Winston. https://specialcollections.nal.usda.gov/speccoll /collectionsguide/darrow/Darrow_TheStrawberry.pdf.

Brulle, Robert J., and David N. Pellow. 2006. "Environmental Justice: Human Health and Environmental Inequalities." *Annual Review of Public Health* 27: 103–124.

Bryant, Bunyan. 1995. "Introduction." In *Environmental Justice: Issues, Policies and Solutions*, edited by Bunyan Bryant. Washington, DC: Island.

Buford, Talia. 2015. "In California, an Unsatisfying Settlement on Pesticide-Spraying." Center for Public Integrity, August 11, 2015. https://publicintegrity.org/environment/in-california -an-unsatisfying-settlement-on-pesticide-spraying/.

Burawoy, Michael. 1979. *Manufacturing Consent.* Chicago: University of Chicago Press.

———. 1976. "The Functions and Reproduction of Migrant Labor: Comparative Material from Southern Africa and the United States." *American Journal of Sociology* 81, no. 5: 1050–1087.

Burdge, Lucy. 2018. "Tom Brady Explains His Hatred of Strawberries on NPR." WEEI Sports Radio, March 19, 2018. https://weei.radio.com/blogs/lucy-burdge/tom-brady-explains -his-hatred-strawberries-npr.

Burke, Brian, and Boone Shear. 2014. "Introduction: Engaged Scholarship for Non-Capitalist Political Ecologies." *Journal of Political Ecology* 21, no. 1: 127–144.

BW Research Partnership. 2018. *2018 Industry, Economic & Workforce Research of Santa Cruz County.* Santa Cruz, CA: Santa Cruz County Workforce Development Board. https:// www.co.santa-cruz.ca.us/portals/0/SCWDB%202018%20Report.pdf.

CA DPR (California Department of Pesticide Regulation). 2018. "Table 4: Pounds of Pesticide Active Ingredients, 1998–2018, by General Use Categories." https://www.cdpr.ca.gov/docs/pur/pur17rep/tables/table4.htm.

———. 2017a. *A Guide to Pesticide Regulation in California: 2017 Update.* Sacramento, CA: CA DPR. https://www.cdpr.ca.gov/docs/pressrls/dprguide/dprguide.pdf.

———. 2017b. *Summary of Pesticide Use Report Data—2017.* Sacramento, CA: CA DPR. https://www.cdpr.ca.gov/docs/pur/pur17rep/17sum.htm#pestuse.

———. 2017c. Data Summary: Pesticide Use in California. Sacramento, CA: CA DPR. Table. https://www.cdpr.ca.gov/docs/pur/pur17rep/17sum.htm#pestuse

———. 2017d. "Table 31: Strawberry." Web page, accessed September 20, 2020. https://www.cdpr.ca.gov/docs/pur/pur17rep/tables/table31.htm.

———. n.d. "California Pesticide Illness Query (CalPIQ)," data for 2016. Web application/database, accessed September 7, 2020. https://apps.cdpr.ca.gov/calpiq/calpiq_input.cfm.

Cabrera, Nolan L., and James O. Leckie. 2009. "Pesticide Risk Communication, Risk Perception, and Self-Protective Behaviors among Farmworkers in California's Salinas Valley." *Hispanic Journal of Behavioral Sciences* 31, no. 2: 258–272.

Cagle, Susie. 2019. "'Bees, Not Refugees': The Environmentalist Roots of Anti-immigrant Bigotry." *Guardian*, August 15, 2019. https://www.theguardian.com/environment/2019/aug/15/anti.

California Federation of Teachers. 2011. *Convention 2011 Resolutions Committee Report: Resolutions and Constitutional Amendments Passed by Delegates to CFT Convention March 18–20 in Manhattan Beach.* Burbank: California Federation of Teachers. https://www.cft.org/sites/main/files/file-attachments/2011_conv_resolutions_final2_jg.pdf?1547665991.

California Strawberry Commission. n.d.-a. "Automation." Web page, accessed September 7, 2020. http://www.calstrawberry.com/en-us/Automation.

———. n.d.-b. "California Export Reports." Web database, accessed September 7, 2020. http://www.calstrawberry.com/en-us/market-data/california-exports.

———. n.d.-c. "8-a-Day." Web page, accessed September 7, 2020. http://www.californiastrawberries.com/welcome/8-a-day/.

Californians for Pesticide Reform. 2014. *Protecting Their Potential: Ensuring California's School Children Are Safe from Hazardous Pesticides.* Sacramento: Californians for Pesticide Reform. https://www.panna.org/sites/default/files/ProtectingTheirPotentialApril2014FINAL.pdf.

Campbell, Monica. 2019. "He Sued His Boss for Violating Employment Rights, Then He Was Deported." *The Week*, September 22, 2019. https://theweek.com/articles/861471/sued-boss-violating-employment-rights-deported.

Cantú, Francisco. 2018. *The Line Becomes a River: Dispatches from the Border.* New York: Penguin Random House.

Carney, Megan A. 2015. *The Unending Hunger: Tracing Women and Food Insecurity across Borders.* Berkeley: University of California Press.

Carson, Rachel. 1962. *Silent Spring.* Boston: Houghton Mifflin.

Cartwright, Elizabeth. 2013. "Eco-Risk and the Case of Fracking." In *Cultures of Energy*, edited by Sarah Strauss, Stephanie Rupp, and Thomas Love, 201–212. Walnut Creek, CA: Left Coast.

———. 2011. "Immigrant Dreams: Legal Pathologies and Structural Vulnerabilities along the Immigration Continuum." *Medical Anthropology* 30, no. 5: 475–495.

Cartwright, Elizabeth, and Lenore Manderson. 2011. "Diagnosing the Structure: Immigrant Vulnerabilities in a Global Perspective." *Medical Anthropology* 30, no. 5: 451–453.

Castañeda, Heide. 2010. "Im/migration and Health: Conceptual, Methodological, and Theoretical Propositions for Applied Anthropology." *NAPA Bulletin* 34, no. 1: 6–17.

Castañeda, Xóchitl, and Patricia Zavella. 2003. "Changing Constructions of Sexuality and Risk: Migrant Mexican Women Farmworkers in California." *Journal of Latin American Anthropology* 8, no. 2: 2–27.

Castillo, Andrea. 2016. "Clovis Wins $22 Million against Shell Oil over Toxic Drinking Water." *Fresno (CA) Bee*, December 21, 2016. https://www.fresnobee.com/news/local/article1222 57349.html.

Castillo, Carla G. 2018. "What the Doctors Don't See: Physicians as Gatekeepers, Injured Latino Immigrants, and Workers' Compensation System." *Anthropology of Work Review* 39, no. 2: 94–104.

CDC (Centers for Disease Control and Prevention). 2019. *National Report on Human Exposure to Environmental Chemicals*. Atlanta: Centers for Disease Control and Prevention. https://www.cdc.gov/exposurereport/index.html.

Cha, Paulette. 2019. "Immigrants and Health in California." Public Policy Institute of California. https://www.ppic.org/publication/immigrants-and-health-in-california/.

Chang, Grace. 2000. *Disposable Domestics: Immigrant Women Workers in the Global Economy*. Cambridge, MA: South End Press.

Charles, Dan. 2012. "The Secret Life of California's World-Class Strawberries." *The Salt*, National Public Radio, May 17, 2012. https://www.npr.org/sections/thesalt/2012/05/17 /152522900/the-secret-life-of-californias-world-class-strawberries.

Chatterjee, Piya, and Sunaina Maira. 2014. *The Imperial University: Academic Repression and Scholarly Dissent*. Minneapolis: University of Minnesota Press.

Chávez, Leo R. 2014. "A Glass Half Empty: Latina Reproduction and Public Discourse." *Human Organization* 63, no. 2: 173–188.

———. 2013. *The Latino Threat: Constructing Immigrants, Citizens and the Nation*. 2nd ed. Stanford, CA: Stanford University Press.

———. 2003. "Immigration Reform and Nativism: The Nationalist Response to the Transnationalist Challenge." In *Perspectives on Las Américas: A Reader in Culture, History, and Representation*, edited by Matthew C. Gutmann, Félix V. Matos Rodríguez, Lynn Stephen, and Patricia Zavella, 418–429. Malden, MA: Blackwell.

Checker, Melissa. 2009. "Practicing Anthropology in 2008: Significant Impacts, Emerging Trends." *American Anthropologist* 111, no. 1: 162–169.

———. 2005. *Polluted Promises: Environmental Racism and the Search for Justice in a Southern Town*. New York: New York University Press.

Chin, Elizabeth. 2013. "The Neoliberal Institutional Review Board, or Why Just Fixing the Rules Won't Help Feminist (Activist) Ethnographers." In *Feminist Activist Ethnography: Counterpoints to Neoliberalism in North America*, edited by Christa Craven and Dána-Ain Davis, 201–216. Lanham, MD: Lexington Books.

Chiu, Yu-Han, Paige L. Williams, Linda Mínguez-Alcarón, Matthew Gillman, Qi Sun, Maria Ospina, Antonia M. Calafat, et al. 2018. "Comparison of Questionnaire-Based Estimation of Pesticide Residue Intake from Fruits and Vegetables with Urinary Concentrations of Pesticide Biomarkers." *Journal of Exposure Science and Environmental Epidemiology* 28, no. 1: 31–39.

Chollett, Donna. 2000. "Neoliberalism's Elusive Benefits: A Case Study of Puruarán, Michoacán." Bulletin of investigation, Universidad Obrera Mexicana.

———. 1996. "Culture, Ideology, and Community: The Dynamics of Accommodation and Resistance to Transformations in the Mexican Sugar Sector." *Culture and Agriculture* 18, no. 3: 98–109.

Chomsky, Aviva. 2007. *"They Take Our Jobs!" And 20 Other Myths about Immigration*. Boston: Beacon.

CIRS (California Institute for Rural Studies). 2018. "Farmworker Housing Study and Action Plan for Salinas Valley and Pájaro Valley." Davis, CA: CIRS. https://www.co.monterey.ca.us/home/showdocument?id=63729.

Clarke, Kamari M. 2010. "Toward a Critically Engaged Ethnographic Practice." *Current Anthropology* 51, no. 2: S301–S312.

Clary, Tim, and Beate Ritz. 2003. "Pancreatic Cancer Mortality and Pesticide Use in California." *American Journal of Industrial Medicine* 43, no. 3: 306–313.

Clean Water Action. 2016. "TCP in California's Drinking Water." Clean Water Action, March 16, 2016. https://www.cleanwateraction.org/features/tcp-californias-drinking-water.

Clifford, James, and George E. Marcus. 1986. "Contemporary Problems of Ethnography in the Modern World System." In *Writing Culture: Experiments in Contemporary Anthropology*, edited by James Clifford and George E. Marcus, 165–193. Berkeley: University of California Press.

Colbert, Stephen. 2010. "Fallback Position—Migrant Worker Pt. 2." *The Colbert Report*, Comedy Central, September 23, 2010. http://www.cc.com/video-clips/puxqvp/the-colbert-report-fallback-position—migrant-worker-pt-2.

Colborn, Theo. 2006. "A Case for Revisiting the Safety of Pesticides: A Closer Look at Neurodevelopment." *Environmental Health Perspectives* 114, no. 1: 10–17.

Colborn, Theo, and Lynn E. Carroll. 2007. "Pesticides, Sexual Development, Reproduction, and Fertility: Current Perspective and Future Direction." *Human and Ecological Risk Assessment: An International Journal* 13, no. 5: 1078–1110.

Colborn, Theo, Dianne Dumanowski, and John Peterson Myers. 1997. *Our Stolen Future: Are We Threatening Our Fertility, Our Intelligence and Survival? A Scientific Detective Story.* New York: Penguin.

Cone, Marta. 2007. "EPA Approves New Pesticide despite Scientists' Concerns." *Los Angeles Times*, October 6, 2007. http://nemaplex.ucdavis.edu/General/News/Methyl Iodide2.htm.

Copeland, Nick, and Christine Labuski. 2013. *The World of Wal-Mart: Discounting the American Dream.* New York: Routledge, 2013.

Cornell, Emily. 2019. "A 'Blue Card' Can Provide Protection for Farm Workers." FoodTank, May 2019. https://foodtank.com/news/2019/05/a-blue-card-can-provide-protection-for-farm-workers/.

Costa, Daniel. 2019. "Employers Increase Their Profits and Put Downward Pressure on Wages and Labor Standards by Exploiting Migrant Workers." Economic Policy Institute, August 27, 2019. https://www.epi.org/publication/labor-day-2019-immigration-policy/.

County of Santa Cruz. 2017. *2017 Crop Report.* Watsonville, CA: County of Santa Cruz, Office of the Agricultural Commissioner. https://www.agdept.com/Portals/10/pdf/CR%202017%20-%20webpage%20copy.pdf?ver=2018-10-17-091626-757&timestamp=1539793319091.

Cox, Shanna, Amanda Sue Niskar, K. M. Venkat Narayan, and Michele Marcus. 2007. "Prevalence of Self-Reported Diabetes and Exposure to Organochlorine Pesticides among Mexican Americans: Hispanic Health and Nutrition Examination Survey, 1982–1984." *Environmental Health Perspectives* 115, no. 2: 1747–1752.

Cranor, Carl F. 2017. *Toxic Failures: How and Why We Are Harmed by Toxic Chemicals.* Oxford: Oxford University Press.

CTPP (Colectivo Todo Poder al Pueblo). 2015. "Farmworkers' Daughter Rejects 'Strawberry Scholarship" in Solidarity with San Quintin." CTPP, May 18, 2015. https://todopoderalpueblo.org/2015/05/19/farmworkers-daughter-rejects-strawberry-scholarship-in-solidarity-with-san-quintin/.

Cullen, Lisa. 2002. *A Job to Die For: Why So Many Americans Are Killed, Injured or Made Ill at Work and What to Do about It*. Monroe, ME: Common Courage.

Cusick, Joan. n.d. "Melissa Dennis: Ohlone Elementary." *Caught in the Drift* (blog), accessed September 7, 2020. https://www.caughtinthedrift.com/Families/Melissa-Dennis.

D'Andrade, Roy. 1995. "Moral Models in Anthropology." *Current Anthropology* 36, no. 3: 399–408.

Darbre, Philippa D. 2017. "Endocrine Disruptors and Obesity." *Current Obesity Research* 6, no. 1: 18–27.

Daria, James. 2019. "Jornalero: Indigenous Migrant Farmworkers along the U.S./Mexican Border." PhD diss., University of Oregon.

Darrow, George M. 1966. *The Strawberry: History, Breeding, and Physiology*. New York: Holt, Rinehart and Winston.

Das, Anindya. 2011. "Farmers' Suicide in India: Implications for Public Mental Health." *International Journal of Social Psychiatry* 57, no. 1: 21–29.

Das, Veena. 1997. "Sufferings, Theodicies, Disciplinary Practices, Appropriations." *International Social Science Journal* 49, no. 154: 563–572.

Davidson, Alan. 2014. "Strawberry." In *The Oxford Companion to Food*, ed. Tom Jaine, 781. Oxford: Oxford University Press.

Davies, Thom. 2019. "Slow Violence and Toxic Geographies: 'Out of Sight' to Whom?" *Environment and Planning C: Politics and Space*, April 19, 2019. https://doi.org/10.1177/2399654419841063.

Davis, Dana-Ain. 2006. "Knowledge in the Service of a Vision." In *Engaged Observer: Anthropology, Advocacy, and Activism*, edited by Victoria Sanford and Asale Angel-Ajani, 228–238. New Brunswick, NJ: Rutgers University Press.

De Leon, Jason. 2015. *The Land of Open Graves: Living and Dying on the Migrant Trail*. Berkeley: University of California Press.

De Long, Nicole E., and Alison C. Holloway. 2017. "Early-Life Chemical Exposures and Risk of Metabolic Syndrome." *Diabetes, Metabolic Syndrome and Obesity* 21, no. 10: 101–109.

Deloria, Vine. 1988. *Custer Died for Your Sins: An Indian Manifesto*. Norman: University of Oklahoma Press.

Dennis, Brady. 2017. "EPA Chief, Rejecting Agency's Own Analysis, Declines to Ban Pesticide despite Health Concerns." *Washington Post*, March 29, 2017. https://www.washingtonpost.com/news/energy-environment/wp/2017/03/29/trump-epa-declines-to-ban-pesticide-that-obama-had-proposed-outlawing/.

Dennis, Brady, and Juliet Eilperin. 2019. "California to Ban Controversial Pesticide, Citing Effects on Child Brain Development." *Washington Post*, May 8, 2019. https://www.washingtonpost.com/climate-environment/2019/05/08/california-ban-controversial-pesticide-citing-effect-child-brain-development/.

Dewan, Pooja, Vikas Jain, Piyush Gupta, and Basu Dev Banerjee. 2013. "Organochlorine Pesticide Residues in Maternal Blood, Cord Blood and Breastmilk and Their Relation to Birth Size." *Chemosphere* 90, no. 5: 1704–1710.

Dilger, Hansjörg, Susann Huschke, and Dominik Mattes. 2015. "Ethics, Epistemology and Engagement: Encountering Values in Medical Anthropology." *Medical Anthropology* 34, no. 1: 1–10.

Doughty, Paul L. 2012. "Back to the Future, Again: From Community Development to PAR." In *Expanding American Anthropology: 1945–1980*, edited by Alice B. Kehoe and Paul L. Doughty, 74–87. Tuscaloosa: University of Alabama Press.

Dowdall, Courtney Marie, and Ryan J. Klotz. 2014. *Pesticides and Global Health: Understanding Agrochemical Dependence and Investing in Sustainable Solutions*. New York: Routledge.

Driscoll's, Inc. n.d. "Joy Starts Here." Web page, accessed September 7, 2020. https://www
.driscolls.com/about/joy-makers.

Duncan, Whitney. 2018. "Acompañamiento/Accompaniment." Society for Cultural Anthropology, January 31, 2018. https://culanth.org/fieldsights/acompañamiento-accompaniment.

DuPuis, E. Melanie. 2002. *Nature's Perfect Food: How Milk Became America's Drink.* New York:
New York University Press.

Durham, Elizabeth. 2016. "The 'Good-Enough Anthropologist.'" *Somatosphere* (blog),
August 26, 2016. http://somatosphere.net/2016/the-good-enough-anthropologist.html/.

Ehrenreich, Barbara. 2010. *Bright-Sided: How Positive Thinking Is Undermining America.* New
York: Metropolitan Books.

Eilperin, Juliet, and Brady Dennis. 2017. "EPA Is Taking More Advice from Industry—And
Ignoring Its Own Scientists." *Washington Post,* November 10, 2017. https://www.washing
tonpost.com/politics/epa-is-taking-more-advice-from-industry--and-ignoring-its-own
-scientists/2017/11/10/aa1fbaba-b8fb-11e7-9e58-e6288544af98_story.html.

Eskenazi, Brenda, Karen Huen, Amy Marks, Kim G. Harley, Asa Bradman, Dana B. Barr, and
Nina Holland. 2010. "PON1 and Neurodevelopment in Children from the CHAMACOS
Study Exposed to Organophosphate Pesticides in Utero." *Environmental Health Perspec-
tives* 118, no. 12: 1775–1781.

Eskenazi, Brenda, Katherine Kogut, Karen Huan, Kim G. Harley, Maryse Bouchard, Asa
Bradman, Dana Boyd-Barr, Caroline Johnson, and Nina Holland. 2014. "Organophosphate Pesticide Exposure, PON1 and Neurodevelopment in School-Age Children from
the CHAMACOS Study." *Environmental Research* 134: 49–57.

Eskenazi, Brenda, Amy R. Marks, Asa Bradman, Laura Fenster, Caroline Johnson, Dana B.
Barr, and Nicholas P. Jewell. 2006. "In Utero Exposure to Dichlorodiphenyltrichloroethane (DDT) and Dichlorodiphenyldichloroethylene (DDE) and Neurodevelopment
among Young Mexican American Children." *Pediatrics* 118, no. 1: 233–241.

Environmental Working Group. 2019. "Pesticides + Poison Gases = Cheap, Year-Round
Strawberries." March 20, 2019. https://www.ewg.org/foodnews/strawberries.php.

———. 1997. "Air Monitoring Detects High Levels of Methyl Bromide Near Elementary
School in Watsonville." Policy memorandum, November 17, 1997. https://static.ewg.org
/files/salsipuedes.pdf?_ga=2.48626085.289436251.1546316780-721688897.1544746330.

EWG Science Team. 2020. "EWG's 2018 Shopper's Guide to Pesticides in Produce." Environmental Working Group. March 25, 2020. https://www.ewg.org/foodnews/summary
.php#dirty-dozen-plus.

Farmer, Paul. 2011. "Accompaniment as Policy." Commencement address, Kennedy School of
Government, Harvard University, May 25, 2011. https://www.lessonsfromhaiti.org/press
-and-media/transcripts/accompaniment-as-policy/.

———. 2004. "An Anthropology of Structural Violence." *Current Anthropology* 45, no. 3:
305–317.

———. 2003. *Pathologies of Power: Health, Human Rights and the New War on the Poor.* Berkeley: University of California Press.

Farmview. n.d. "The Farmland Monitoring Project." Web page, accessed September 7, 2020.
https://farmview.herokuapp.com/about/.

Farmworker Justice. 2019. Selected Statistics on Farmworkers 2015–2016. Washington, DC:
Farmworker Justice. Accessed September 7, 2020. http://www.farmworkerjustice.org/wp
-content/uploads/2019/05/NAWS-Data-FactSheet-05-13-2019-final.pdf.

———. 2013. *Exposed and Ignored: How Pesticides Are Endangering Our Nation's Farmworkers.*
Washington, DC: Farmworker Justice. https://kresge.org/sites/default/files/Exposed
-and-ignored-Farmworker-Justice-KF.pdf.

Farquhar, Stephanie, Julie Samples, Santiago Ventura, Shelley Davis, Michelle Abernathy, Linda McCaulty, Nancy N. Cuilwick, and Nargass Shadbeh. 2008. "Promoting the Occupational Health of Indigenous Farmworkers." *Journal of Immigrant and Minority Health* 10, no. 3: 269–280.

Fathallah, Fadi A. 2010. "Musculoskeletal Disorders in Labor-Intensive Agriculture." *Applied Ergonomics* 41, no. 6: 738–743.

FELS (Farm Employers Labor Service). 2014 *Summary of Employment Requirements for California Winegrape Growers*. 2014 ed. Sacramento, CA: FELS / Vineyard Team. http://www.vineyardteam.org/files/resources/Summary-of-Employment_2014.pdf.

Ferrar, Kyle. 2017. "Tracking Refinery Emissions in California's Bay Area Refinery Corridor." FracTracker Alliance, May 10, 2017. https://www.fractracker.org/2017/05/tracking-refinery-emissions-in-californias-bay-area-refinery-corridor/.

Figueroa Sanchez, Teresa. 2013. "California Strawberries: Mexican Immigrant Women Sharecroppers, Labor and Discipline." *Anthropology of Work Review* 34, no. 1: 15–26.

Fisher, Andrew. 2018. *Big Hunger: The Unholy Alliance between Corporate American and Antihunger Groups*. Cambridge, MA: MIT Press.

Fischer, Edward F., and Peter Benson. 2006. *Broccoli and Desire: Global Connections and Maya Struggles in Postwar Guatemala*. Stanford, CA: Stanford University Press.

Fjord, Lakshmi, and Lenore Manderson. 2009. "Anthropological Perspectives on Disasters and Disability: An Introduction." *Human Organization* 68, no. 1: 64–72.

Fletcher, Stevenson Whitcomb. 1917. *The Strawberry in North America: History, Origin, Botany and Breeding*. New York: Macmillan.

Flocks, Joan, and Paul Monaghan. 2003. "Collaborative Research with Farmworkers in Environmental Justice." *Practicing Anthropology* 25, no. 1: 6–9.

Foley, Douglas E. 1999. "The Fox Project: A Reappraisal." *Current Anthropology* 40, no. 2: 171–192.

Fortun, Kim. 2001. *Advocacy after Bhopal: Environmentalism, Disaster, New Global Orders*. Chicago: University of Chicago Press.

Freidberg, Susanne. 2009. *Fresh: A Perishable History*. Cambridge, MA: Harvard University Press.

Friedland, William H. 1994. "The New Globalization: The Case of Fresh Produce." In *From Columbus to ConAgra: The Globalization of Agriculture and Food*, edited by Alessandro Bonano, Lawrence Busch, William Friedland, Lourdes Gouveia, and Enzo Mingione, 210–231. Lawrence: University Press of Kansas.

———. 1984. "Commodity Systems Analysis: An Approach to the Sociology of Agriculture." In *Research in Rural Sociology and Development*, edited by Harry K. Schwarzweller, 221–235. London: JAI.

Friedland, William H., Amy E. Barton, and Robert J. Thomas. 1981. *Manufacturing Green Gold: Capital, Labor, and Technology in the Lettuce Industry*. Cambridge: Cambridge University Press.

Froines, John, Susan E. Kegley, Timothy F. Malloy, and Sarah Kobylewski. 2013. *Risk and Decision: Evaluating Pesticide Approval in California; Review of the Methyl Iodide Registration Process*. Los Angeles: UCLA Sustainable Technology and Policy Program. https://www.pesticideresearch.com/site/wp-content/uploads/2012/05/Risk_and_Decision_Report_2013.pdf.

Froines, John R., Paul Blanc, Katharine Hammond, Dale Hattis, Ed Loechler, Ron Melnik, Tom McKone, and Theodore Slotkin. 2010. *Report of the Scientific Review Committee on Methyl Iodide to the Department of Pesticide Regulation*. Sacramento: California Department of Pesticide Regulation. https://civileats.com/wp-content/uploads/2011/03/SRC_letter_to_DPR.pdf.

Furlong, Melissa, Caroline M. Tanner, Samuel M. Goldman, Grace S. Bhudhikanok, Aaron Blair, Anabel Chade, Kathleen Comyns, et al. 2015. "Protective Glove Use and Hygiene Habits Modify the Associations of Specific Pesticides with Parkinson's Disease." *Environment International* 75: 144–150.

Gálvez, Alyshia. 2018. *Eating NAFTA: Trade, Food Policies and the Destruction of Mexico.* Berkeley: University of California Press.

———. 2011. *Patient Citizens, Immigrant Mothers.* New Brunswick, NJ: Rutgers University Press.

Gan, Elaine, Anna Tsing, Heather Swanson, and Nils Bubandt. 2017. "Introduction: Haunted Landscapes of the Anthropocene." In *Arts of Living on a Damaged Planet: Ghosts,* edited by Anna Tsing, Heather Swanson, Elaine Gan, and Nils Bubandt, G1–G13. Minneapolis: University of Minnesota Press.

Gao, Suduan, Bradley D. Hanson, Dong Wang, Gregory T. Browne, Ruijun Qin, Husein Ajwa, and Scott R. Yates. 2011. "Methods Evaluated to Minimize Emissions from Preplant Soil Fumigation." *California Agriculture* 65, no. 1: 41–46.

Garcia, Matt. 2014. *From the Jaws of Victory: The Triumph and Tragedy of Cesar Chavez and the Farm Worker Movement.* Berkeley: University of California Press.

Garcia, Robert, Brent Newell, and Madeline Stano. 2013. "Green Justice: Three Generations Sue U.S. EPA over Toxic Pesticides at Schools." KCET, August 29, 2013. https://www.kcet .org/history-society/three-generations-sue-us-epa-over-toxic-pesticides-at-schools.

Garcia, Victor. 2005. "The Mushroom Industry and the Emergence of Mexican Enclaves in Southern Chester County Pennsylvania, 1960–1990." *Journal of Latino / Latin American Studies* 1, no. 4: 67–88.

———. 1997. *Mexican Enclaves in the U.S. Northeast: Immigrant and Migrant Mushroom Workers in Southern Chester County, Pennsylvania.* East Lansing: Julian Samora Research Institute, Michigan State University. https://jsri.msu.edu/upload/research-reports/rr27.pdf.

García, Victor, and Edward Gondolf. 2004. "Transnational Mexican Farmworkers and Problem Drinking: A Review of the Literature." *Contemporary Drug Problems* 31, no. 1: 129–161.

Gareau, Brian J. 2013. *From Precaution to Profit: Contemporary Challenges to Environmental Protection in the Montreal Protocol.* New Haven, CT: Yale University Press.

Gaspar, Fraser W., Kim G. Harley, Katherine Kogut, Jonathan Chevrier, Ana Maria Mora, Andreas Sjodin, and Brenda Eskenazi. 2015. "Prenatal DDT and DDE Exposure and Child IQ in the CHAMACOS Cohort." *Environmental International* 85: 206–212.

Gates, Marilyn. 1993. *In Default: Peasants, the Debt Crisis and the Agricultural Challenge in Mexico.* Boulder, CO: Westview.

Gemmill, Allison, Robert B. Gunier, Asa Bradman, Brenda Eskenazi, and Kim G. Harley. 2013. "Residential Proximity to Methyl Bromide Use and Birth Outcomes in an Agricultural Population in California." *Environmental Health Perspectives* 121, no. 6: 737–743.

Getz, Christy, Sandy Brown, and Aimee Shreck. 2008. "Class Politics and Agricultural Exceptionalism in California's Organic Agriculture Movement." *Politics and Society* 36, no. 4: 478–507.

Gibson, Walter S. 2003. "The Strawberries of Hieronymus Bosch." *Cleveland Studies in the History of Art* 8: 24–33.

Gilmore, Ruth Wilson. 2008. "Forgotten Places and the Seeds of Grassroots Planning." In *Engaging Contradictions: Theory, Politics and Methods of Activist Scholarship,* edited by Charles R. Hale and Craig Calhoun, 31–61. Berkeley: University of California Press.

Giulivo, Monica, Miren López de Alda, Ettore Capri, and Damia Barceló. 2016. "Human Exposure to Endocrine Disrupting Compounds: Their Role in Reproductive Systems, Metabolic Syndrome and Breast Cancer. A Review." *Environmental Research* 151: 251–264.

Gledhill, John. 1999. "Official Masks and Shadow Powers: Towards an Anthropology of the Dark Side of the State." *Urban Anthropology and Studies of Cultural Systems and World Economic Development* 28, nos. 3–4: 199–251.

———. 1998. "The Mexican Contribution to Restructuring US Capitalism: NAFTA as an Instrument of Flexible Accumulation." *Critique of Anthropology* 18, no. 3: 279–296.

Goizueta, Roberto. 2009. *Christ Our Companion: Toward a Theological Aesthetics of Liberation.* Maryknoll, NY: Orbis Books.

Goldberg, Ted. 2018a. "Company Won't Pay More Than $5,000 after Pesticide Exposure Sickens 17 Farmworkers." KQED, July 31, 2018. https://www.kqed.org/news/11681690/2017 -monterey-county-pesticide-drift-incident-near-salinas.

———. 2018b. "Workers in Central Coast Pesticide Drift Tied to Dole, Driscoll's Were Sick for Days." KQED, July 5, 2018. https://www.kqed.org/news/11678534/workers-in-central -coast-pesticide-drift-tied-to-dole-driscolls-were-sick-for-days.

———. 2017. "Firms Fined for Pesticide Incident That Sickened 92 in Bakersfield." KQED, December 18, 2017. https://www.kqed.org/news/11637296/firms-fined-for-pesticide -incident-that-sickened-92-near-bakersfield.

Goldman, Lisa, Brenda Eskenazi, Asa Bradman, and Nicholas P. Jewell. 2004. "Risk Behaviors for Pesticide Exposure among Pregnant Women Living in Farmworker Households in Salinas, California." *American Journal of Industrial Medicine* 45, no. 6: 491–499.

Goldstein, Donna. 2017. "Invisible Harm: Science, Subjectivity and the Things We Cannot See." *Culture, Theory and Critique* 58, no. 4: 321–329.

Gong, Jake. 2015. "Are Immigrants to Blame for California's Drought?" *The Highlander* (University of California–Riverside), June 23, 2015. https://www.highlandernews.org/20183 /are-immigrants-to-blame-for-californias-drought/.

González, Roberto J., and Rachael Stryker. 2014. "On Studying Up, Down and Sideways: What's at Stake?" In *Up, Down and Sideways: Anthropologists Trace the Pathways of Power,* edited by Rachael Stryker and Roberto J. González, 1–26. New York: Berghahn Books.

Gonzalez-Barrera, Ana. 2015. "More Mexicans Leaving Than Coming to the U.S." Pew Research Center, November 19, 2015. https://www.pewresearch.org/hispanic/2015/11/19 /more-mexicans-leaving-than-coming-to-the-u-s/.

Goodyear, Dana. 2017. "How Driscoll's Reinvented the Strawberry." *New Yorker,* August 14, 2017. https://www.newyorker.com/magazine/2017/08/21/how-driscolls-reinvented-the -strawberry.

Gore, Andrea C., Vesna A. Chappell, Suzanne E. Fenton, Jodi A. Flaws, Angel Nadal, Gail S. Prins, Jussi Toppari, and R. Thomas Zoeller. 2015. "EDC-2: The Endocrine Society's Second Scientific Statement on Endocrine-Disrupting Chemicals." *Endocrine Reviews* 36, no. 6: E1–E50.

Grabell, Michael. 2017. "Sold for Parts." ProPublica, March 1, 2017. https://www.propublica .org/article/case-farms-chicken-industry-immigrant-workers-and-american-labor-law.

———. 2015a. "The Fallout of Workers' Comp 'Reforms': 5 Tales of Harm." ProPublica, March 25, 2015. https://www.propublica.org/article/workers-compensation-injured-workers-share -%0D%0Astories-of-harm%0D%0A.

———. 2015b. "How Much Is Your Arm Worth? Depends on Where You Work." ProPublica, March 5, 2015. https://www.propublica.org/article/how-much-is-your-arm-worth-depends -where-you-work.

———. 2015c. "'I Try to Forget.'" ProPublica, March 4, 2015. https://www.propublica.org /article/photos-living-through-california-workers-comp-cuts.

———. 2015d. "State Lawmakers to Investigate Workers' Comp Opt Out." ProPublica, November 25, 2015. https://www.propublica.org/article/state-lawmakers-to-investigate -workers-comp-opt-out.

———. 2015e. "Tyson Food's Secret Recipe for Carving Up Workers' Comp." ProPublica, December 11, 2015. https://www.propublica.org/article/tyson-foods-secret-recipe-for -carving-up-workers-comp.

Grabell, Michael, and Howard Berkes. 2017. "They Got Hurt at Work. Then They Got Deported." ProPublica, August 16, 2017. https://www.propublica.org/article/they-got -hurt-at-work-then-they-got-deported.

Grabell, Michael, and Lena V. Groeger. 2015. "Methodology for Workers' Comp Benefits: How Much Is a Limb Worth?" ProPublica, March 5, 2015. https://www.propublica.org /article/workers-comp-benefits-how-much-is-a-limb-worth-methodology.

Grandjean, Phillippe. 2016. "Paracelsus Revisited: The Dose Concept in a Complex World." *Basic Clinical Pharmacological Toxicology* 119, no. 2: 126–132.

Gray, Margaret. 2013. *Labor and the Locavore: The Making of a Comprehensive Food Ethic.* Berkeley: University of California Press.

Greenbaum, Susan D. 2015. *Blaming the Poor: The Long Shadow of the Moynihan Report on Cruel Images about Poverty.* New Brunswick, NJ: Rutgers University Press.

Greenfield, Victoria, Blas Nuñez-Neto, Ian Mitch, Joseph C. Chang, and Etienne Rosas. 2019. *Human Smuggling and Associated Revenues: What Do or Can We Know About Routes from Central America to the United States?* Santa Monica, CA: RAND Corporation Homeland Security Operational Analysis Center. https://www.rand.org/content/dam/rand/pubs /research_reports/RR2800/RR2852/RAND_RR2852.pdf.

Griffith, David. 2009. "Unions without Borders: Organizing and Enlightening Immigrant Farm Workers." *Anthropology of Work Review* 30, no. 2: 54–66.

Gross, Liza. 2019. "More Than 90 Percent of Americans Have Pesticides or Their Byproducts in Their Bodies." *The Nation*, March 21, 2019. https://www.thenation.com/article /pesticides-farmworkers-agriculture/.

Gross, Rachel. 2011. "Farmers Seek to Raise Standards for Berries." *New York Times*, September 23, 2011. https://www.nytimes.com/2011/09/23/us/farmers-seek-to-raise-standards -for-berries.html.

Grossman, Elizabeth. 2014. "Banned in Europe, Safe in the U.S." Ensia, June 9, 2014. https:// ensia.com/features/banned-in-europe-safe-in-the-u-s/.

Guillette, Elizabeth, Maria. Mercedes Meza, Maria Guadalupe Aguilar, Alma Delia Soto and Idalia Enedina Garcia. 1998. "An Anthropological Approach to the Evaluation of Preschool Children Exposed to Pesticides in Mexico." *Environmental Health Perspectives* 106, no. 6: 347–353.

Gupta, Akhil. 2012. *Red Tape: Bureaucracy, Structural Violence and Poverty in India.* Durham, NC: Duke University Press.

Guthman, Julie. 2019. *Wilted: Pathogens, Chemicals, and the Fragile Future of the Strawberry Industry.* Berkeley: University of California Press.

———. 2018. "Strawberry Fields Forever? When Soil Muddies Sustainability." *The Breakthrough* 9 (June 25, 2018). https://thebreakthrough.org/journal/no-9-summer-2018/strawberry -fields-forever.

———. 2017. "Lives versus Livelihoods? Deepening the Regulatory Debates on Soil Fumigants in California's Strawberry Industry." *Antipode* 49, no. 1: 86–105.

———. 2011. *Weighing In: Obesity, Food Justice, and the Limits of Capitalism.* Berkeley: University of California Press.

————. 2009. "Unveiling the Unveiling: Commodity Chains, Commodity Fetishism and the 'Value' of Voluntary, Ethical Food Labels." In *Frontiers of Commodity Chain Research*, edited by Jennifer Bair, 190–206. Stanford, CA: Stanford University Press.

————. 2008. "Neoliberalism and the Making of Food Politics in California." *Geoforum* 39, no. 3: 1171–1183.

————. 2004. *Agrarian Dreams: The Paradox of Organic Farming in California*. Berkeley: University of California Press.

Guthman, Julie, and Sandy Brown. 2017. "How Midas Lost Its Golden Touch: Neoliberalism and Activist Strategy in the Demise of Methyl Iodide in California." In *The New Food Activism: Opposition, Cooperation and Collective Action*, edited by Alison Alkon and Julie Guthman, 80–105. Berkeley: University of California Press.

————. 2016a. "'I Will Never Eat Another Strawberry Again!': The Biopolitics of Consumer-Citizenship in the Fight against Methyl Iodide." *Agriculture and Human Values* 33, no. 3: 575–585.

————. 2016b. "Midas' Not-So-Golden Touch: On the Demise of Methyl Iodide as a Soil Fumigant in California." *Journal of Environmental Policy and Planning* 18, no. 3: 324–341.

Gutiérrez, Elena R. 2008. *Fertile Matters: The Politics of Mexican-Origin Women's Reproduction*. Austin: University of Texas Press.

Gutiérrez, Gustavo. 1973. *The Power of the Poor in History: Selected Writings*. Maryknoll, NY: Orbis Books.

Hale, Charles R. 2008. "Introduction." In *Engaging Contradictions: Theory, Politics and Methods of Activist Scholarship*, edited by Charles R. Hale, 1–30. Berkeley: University of California Press.

————. 2006. "Activist Research v. Cultural Critique: Indigenous Land Rights and the Contradictions of Politically Engaged Anthropology." *Cultural Anthropology* 21, no. 1: 96–120.

————. 2001. "What Is Activist Research?" *Social Science Research Council: Items and Issues* 2, nos. 1–2: 13–15.

Hamer, Fannie Lou. 2011. "I'm Sick and Tired of Being Sick and Tired." In *Speeches of Fannie Lou Hamer: To Tell It Like It Is*, edited by Maegan Parker Brooks and Davis W. Houck, 57–64. Jackson: University Press of Mississippi.

Han, Clara. 2012. *Life in Debt: Times of Care and Violence in Neoliberal Chile*. Berkeley: University of California Press.

Hanna-Attisha, Mona. 2018. *What the Eyes Don't See: A Story of Crisis, Resistance, and Hope in an American City*. New York: Penguin Random House.

Harper, Janice. 2004. "Breathless in Houston: A Political Ecology of Health Approach to Understanding Environmental Health Concerns." *Medical Anthropology* 23, no. 4: 295–326.

Harrison, Faye V., ed. 2010. *Decolonizing Anthropology: Moving Further toward an Anthropology of Liberation*. 2nd ed. Arlington, VA: Association of Black Anthropologists / American Anthropological Association.

Harrison, Jill Lindsey. 2011. *Pesticide Drift and the Pursuit of Environmental Justice*. Cambridge, MA: MIT Press.

————. 2008. "Abandoned Bodies and Spaces of Sacrifice: Pesticide Drift Activism and the Contestation of Neoliberal Environmental Politics in California." *Geoforum* 39, no. 3: 1197–1214.

————. 2006. "'Accidents' and Invisibilities: Scaled Discourse and the Naturalization of Regulatory Neglect in California's Pesticide Drift Conflict." *Political Geography* 25, no. 5: 506–529.

Hartwick, Elaine. 1998. "Geographies of Consumption: A Commodity-Chain Approach." *Environment and Planning D: Society and Space* 16, no. 4: 423–437.

Hawkes, Glenn R., and Martha C. Stiles. 1986. "Attitudes about Pesticide Safety." *California Agriculture*, May–June 1986, 19–22.

Heidbrink, Lauren. 2018. "Care in Contexts of Child Detention." Society for Cultural Anthropology, January 31, 2018. https://culanth.org/fieldsights/care-in-contexts-of-child-detention.

Held, Lisa. 2018. "Why Food Insecurity Is a Global Farmworker Issue." Civil Eats, December 21, 2018. https://civileats.com/2018/12/21/why-food-insecurity-is-a-global-farmworker-issue/.

Henke, Christopher R. 2008. *Cultivating Science, Harvesting Power: Science and Industrial Agriculture in California*. Cambridge, MA: MIT Press.

Henner, Mishka. 2015. "Op-Ed: How the Meat Industry Marks the Land—In Pictures." *Los Angeles Times*, December 27, 2015. https://www.latimes.com/opinion/op-ed/la-oe-marks-on-the-land-html-20151222-htmlstory.html.

Herrera, Jack. 2018. "As Wildfire Smoke Fills the Air, Farmworkers Continue to Labor in the Fields." *Pacific Standard*, November 14, 2018. https://psmag.com/environment/as-wildfire-smoke-fills-the-air-farmworkers-continue-to-labor-in-the-fields.

Hetchman, Kevin. 2020. "Farmland Goes up in Smoke as Fires Rage." *Daily Democrat*, August 27, 2020. https://www.dailydemocrat.com/2020/08/27/farmland-goes-up-in-smoke-as-fires-rage/

Hetherington, Kregg. 2016. "When Plants Farm Themselves." Society for Cultural Anthropology, July 26, 2016. https://culanth.org/fieldsights/when-plants-farm-themselves

Heyman, Josiah. 2011. "An Academic in an Activist Coalition: Recognizing and Bridging Role Conflicts." *Annals of Anthropological Practice* 35, no. 2: 136–153.

———. 2003. "The Inverse of Power." *Anthropological Theory* 3, no. 2: 139–156.

Hines, Sarah. 2006. "The Bracero Program: 1942–1964." CounterPunch, April 21, 2006. https://www.counterpunch.org/2006/04/21/the-bracero-program-1942-1964-2/.

Holden, Emily. 2019. "Trump EPA Insists Monsanto's Roundup Is Safe, despite Cancer Cases." *Guardian*, April 30, 2019. https://www.theguardian.com/business/2019/apr/30/monsanto-roundup-trump-epa-cancer.

Holmes, Seth M. 2013. *Fresh Fruit, Broken Bodies: Migrant Farmworkers in the United States*. Berkeley: University of California Press.

———. 2011. "Structural Vulnerability and Hierarchies of Ethnicity and Citizenship on the Farm." *Medical Anthropology* 30, no. 4: 425–449.

———. 2007. "Oaxacans Like to Work Bent Over': The Naturalization of Social Suffering among Berry Farm Workers." *International Migration* 45, no. 3: 39–66.

Holtcamp, Wendee. 2012. "Obesogens: An Environmental Link to Obesity." *Environmental Health Perspectives* 120, no. 2: a62–a68.

Holt-Giménez, Eric. 2017a. "Death in the Streets—And the Fields: The Privilege, Promises and Violence of White Supremacy." *Huffington Post*, August 17, 2017. https://www.huffpost.com/entry/death-in-the-streetsand-the-fields-the-privilege_b_59965ebde4b03b5e472cee60.

———. 2017b. *A Foodie's Guide to Capitalism*. New York: Monthly Review Press.

Holtz, Timothy H., Seth M. Holmes, Scott Stonington, and Leon Eisenberg. 2006. "Health Is Still Social: Contemporary Examples in the Age of the Genome." *PLOS Medicine* 3, no. 10: 1663–1666.

Horton, Sarah, Cesar Abadía, Jessica Mulligan, and Jennifer Jo Thompson. 2014. "Critical Anthropology of Global Health 'Takes a Stand' Statement: A Critical Medical Anthropological Approach to the U.S.'s Affordable Care Act." *Medical Anthropology Quarterly* 28, no. 1: 1–22.

Horton, Sarah B. 2016a. "Ghost Workers: The Implications of Governing Immigration through Crime for Migrant Workplaces." *Anthropology of Work Review* 37, no. 1: 11–23.

———. 2016b. *They Leave Their Kidneys in the Fields: Illness, Injury and Illegality among U.S. Farmworkers.* Berkeley: University of California Press.

Howard, Sarah G. 2019. "Exposure to Environmental Chemicals and Type 1 Diabetes: An Update." *Journal of Epidemiology and Community Health* 73, no. 6: 483–488.

Howe, Cymene. 2016. "Negative Space: Unmovement and the Study of Activism." In *Impulse to Act: A New Anthropology of Resistance,* edited by Alexandrakis Oton, 161–182. Bloomington: Indiana University Press.

Huang, Priscilla. 2008. "Anchor Babies, Over-Breeders and the Population Bomb: The Reemergence of Nativism and Population Control in Anti-Immigration Policies." *Harvard Law and Policy Review* 2: 385–406.

Huber, Bridget. 2011. "Head of CA Department of Pesticide Regulation Leaves Post to Work for Chemical Giant." Civil Eats, March 18, 2011. https://civileats.com/2011/03/18/head-of-ca-department-of-pesticide-regulation-leaves-post-to-work-for-chemical-giant/.

Huen, Karen, Paul Yousefi, Kelly Street, Brenda Eskenazi, and Nina Holland. 2015. "PON1 as a Model for Integration of Genetic, Epigenetic and Expression Data on Candidate Susceptibility Genes." *Environmental Epigenetics* 1, no. 1. https://doi.org/10.1093/eep/dvv003.

Husaini, Majad Masood, and Malik Zainul Abdin. 2008. "Development of Transgenic Strawberry (*Fragaria x Ananassa* Duch.) Plants Tolerant to Salt Stress." *Plant Science* 174, no. 4: 446–455.

Huschke, Susann. 2015. "Giving Back: Activist Research with Undocumented Migrants in Berlin." *Medical Anthropology* 34, no. 1: 54–69.

Ibarra, Nicholas. 2019. "Watsonville Latest City to Ban Roundup, Other Glyphosate Weed-Killers." *Santa Cruz (CA) Sentinel,* May 5, 2019. https://www.santacruzsentinel.com/2019/05/05/watsonville-latest-city-to-ban-roundup-other-glyphosate-weed-killers/

———. 2017. "Hundreds Face Layoffs as Dole Closes Watsonville Operations." *Santa Cruz (CA) Sentinel,* September 7, 2017. https://www.santacruzsentinel.com/2017/09/07/hundreds-face-layoffs-as-dole-closes-watsonville-operations/.

Ibekwe, A. M. 2004. "Effects of Fumigants on Non-Target Organisms in Soils." *Advances in Agronomy* 83: 1–35.

Indigenous Action Media. 2014. *Accomplices Not Allies: Abolishing the Ally Industrial Complex.* Flagstaff, AZ: Indigenous Action, 2014. http://www.indigenousaction.org/wp-content/uploads/Accomplices-Not-Allies-print.pdf.

Indigenous Farmworker Study. n.d. "Indigenous Mexicans in California Agriculture." Home page, accessed September 7. 2020. http://www.indigenousfarmworkers.org/.

IndyBay. 2017. "Highly Militarized ICE Raids on California's Central Coast." It's Going Down, February 16, 2017. https://itsgoingdown.org/highly-militarized-ice-raids-californias-central-coast/.

Ingold, Tim. 2014. "That's Enough about Ethnography!" *HAU: Journal of Ethnographic Theory* 4, no. 1: 383–395.

Iserson, Kenneth V., and John C. Moskop. 2007. "Triage in Medicine, Part I: Concept, History and Types." *Annals of Emergency Medicine* 49, no. 3: 275–281.

Jackson, John L., Jr. 2020. "On Ethnographic Sincerity." *Current Anthropology* 51, no. 2: S279–S287.

James, D. J., A. J. Passey, A. D. Webster, D. J. Barbara, P. Viss, A. M. Dandekar, and S. L. Uratsu. 1993. "Transgenic Apples and Strawberries: Advances in Transformation, Introduction of Genes for Insect Resistance and Field Studies of Tissue Cultured Plants." In *Acta Horticulturae,* vol. 336, *II International Symposium on In Vitro Culture and Horticultural Breeding,* edited by F. A. Hammerschlag and L. D. Owens, 179–184. Leuven, Belgium: International Society for Horticultural Science.

Ji, Bu-Tian, Debra T. Silverman, Patricia A. Stewart, Aaron Blair, G. Marie Swanson, Dalsu Baris, Raymond S. Greenberg, et al. 2001. "Occupational Exposure to Pesticides and Pancreatic Cancer." *American Journal of Industrial Medicine* 39, no. 1: 92–99.

Johnson, Jim. 2020. "Farmworkers Facing Multiple Threats with COVID-19, Heat, Wildfire Smoke." *Monterey (CA) Herald,* August 21, 2020. https://www.montereyherald.com/2020/08/21/farmworkers-facing-multiple-threats-with-covid-19-heat-wildfire-smoke/

Johnson, Noor. 2014. "Thinking through Affect: Inuit Knowledge on the Tundra and in Global Environmental Politics." *Journal of Political Ecology* 21, no. 1: 161–177.

Johnston, Barbara Rose. 2010. "Social Responsibility and the Anthropological Citizen." *Current Anthropology* 51, no. 2: S235–S247.

———. 2001a. "Anthropology and Environmental Justice: Analysts, Advocates, Mediators and Troublemakers." In *New Directions in Anthropology and Environment: Intersections,* edited by Carole L. Crumley, 132–149. Walnut Creek, CA: Altamira.

———. 2001b. "Backyard Anthropology and Community-Based Environmental Protection: Lessons from the SfAA Environmental Anthropology Project." *Practicing Anthropology* 23, no. 3: 2–6.

Jolivette, Andrew J. 2015. "Research Justice: Radical Love as a Strategy for Social Change." In *Research Justice: Methodologies for Social Change,* edited by Andrew J. Jolivette, 5–12. Chicago: Policy Press.

Jones, Donna. 2011. "State to Monitor Pesticide Levels near Watsonville School: Methyl Bromide Program Settles 12-Year-Old Civil Rights Complaint." *Santa Cruz (CA) Sentinel,* August 26, 2011. https://www.santacruzsentinel.com/2011/08/26/state-to-monitor-pesticide-levels-near-watsonville-school-methyl-bromide-program-settles-12-year-old-civil-rights-complaint/.

Jones, Liz. 2017. "A Death Pushed These Farmworkers to Protest. Now an Investigation Is Underway." KUOW, August 10, 2017. https://kuow.org/stories/death-pushed-these-farmworkers-protest-now-investigation-underway/.

Kang, Gail A., Jeff M. Bronstein, Donna L. Masterman, Matthew Redelings, Jarrod A. Crum, and Beate Ritz. 2005. "Clinical Characteristics in Early Parkinson's Disease in a Central California Population-Based Study." *Movement Disorders* 20, no. 9: 1133–1142.

Kan-Rice, Pamela. 2008. "UC Strawberries Go to the Olympics." Press release. University of California, Agriculture and Natural Resources, August 6, 2008. https://ucanr.edu/blogs/blogcore/postdetail.cfm?postnum=549.

Karst, Tom. 2018. "Late Season Berries 2018 Business Updates: Dole Looks for Strong Import Program." Produce Market Guide, August 9, 2018. https://www.producemarketguide.com/article/late-season-berries-2018-business-updates.

Kavasch, E. Barrie. 1995. *Enduring Harvests: Native American Foods and Festivals for Every Season.* Old Saybrook, CT: Globe Pequot.

Kehoe, Alice Beck, and Paul L. Doughty. 2012. *Expanding American Anthropology, 1945–1980: A Generation Reflects.* Tuscaloosa: University of Alabama Press.

Kelley, Maureen A., Joan D. Flocks, Jeannie Economos, and Linda A. McCauley. 2013. "Female Farmworkers' Health during Pregnancy: Health Care Providers' Perspectives." *Workplace Health and Safety* 61, no. 7: 308–313.

Khasnabish, Alex. 2016. "Within, against, beyond: The Radical Imagination in the Age of the Slow-Motion Apocalypse." In *Impulse to Act: A New Anthropology of Resistance,* edited by Alexandrakis Oton, 231–245. Bloomington: Indiana University Press.

Khoka, Sasha. 2017. "California Finally Begins Regulating Cancer-Causing Chemical Found in Drinking Water." KQED, July 21, 2017. https://www.kqed.org/science/560344/theres-a-cancer-causing-chemical-in-my-drinking-water-but-california-isnt-regulating-it.

Kivel, Paul. 2017. "Social Service or Social Change?" In *The Revolution Will Not Be Funded: Beyond the Non-Profit Industrial Complex*, edited by INCITE!, 129–150. Durham, NC: Duke University Press.

Kleinman, Arthur, and Peter Benson. 2006. "Anthropology in the Clinic: The Problem of Cultural Competency and How to Fix It." *PLoS Medicine* 3, no. 10: 1673–1676.

Kline, Nolan. 2019. *Pathogenic Policing: Immigration Enforcement and Health in the U.S. South.* New Brunswick, NJ: Rutgers University Press.

———. 2010. "Disparate Power and Disparate Resources: Collaboration between Faith-Based and Activist Organizations for Central Florida Farmworkers." *NAPA Bulletin* 33, no. 8: 126–142.

Kohl-Arenas, Erica. 2016. *The Self-Help Myth: How Philanthropy Fails to Alleviate Poverty.* Berkeley: University of California Press.

Kopytoff, Igor. 1986. "The Cultural Biography of Things: Commoditization as Process." In *The Social Life of Things: Commodities in Cultural Perspective*, edited by Arjun Appadurai, 64–91. Cambridge: Cambridge University Press.

Kosek, Jake. 2006. *Understories: The Political Life of Forests in Northern New Mexico.* Durham, NC: Duke University Press.

Kovach, Margaret. 2009. *Indigenous Methodologies: Characteristics, Conversations, Contexts.* Toronto: University of Toronto Press, 2009.

Krieger, Nancy. 2010. "Workers Are People Too: Societal Aspects of Occupational Health Disparities—An Ecosocial Perspective." *American Journal of Industrial Medicine* 53: 104–115.

———. 2001. "Theories for Social Epidemiology in the 21st Century: An Ecosocial Perspective." *International Journal of Epidemiology* 30, no. 4: 668–677.

———. 1994. "Epidemiology and the Web of Causation: Has Anyone Seen the Spider?" *Social Science and Medicine* 39, no. 7: 887–903.

Kulish, Nicholas, and Mike McIntire. 2019. "The New Nativists: Why an Heiress Spent Her Fortune Trying to Keep Immigrants Out." *New York Times*, August 14, 2019. https://www.nytimes.com/2019/08/14/us/anti-immigration-cordelia-scaife-may.html.

Lamphere, Louise. 2018. "The Transformation of Ethnography: From Malinowski's Tent to the Practice of Collaborative/Activist Anthropology." *Human Organization* 77, no. 1: 64–76.

Lantham, Jonathan. 2019. "The EPA and the Chemical Industry: A Cosy Alliance." *Against the Grain*, KPFA, June 24, 2019. https://kpfa.org/episode/against-the-grain-june-24-2019/.

Lantigua, John. 2011. "Illegal Immigrants Pay Social Security Tax, Won't Benefit." *Seattle Times*, December 28, 2011. https://www.seattletimes.com/nation-world/illegal-immigrants-pay-social-security-tax-wont-benefit/.

Laurienti, Paul J., Jonathan H. Burdette, Jennifer Talton, Carey N. Pope, Phillip Summers, Francis O. Walker, Sara A. Quandt, et al. 2016. "Brain Anatomy in Latino Farmworkers Exposed to Pesticides and Nicotine." *Journal of Occupational and Environmental Medicine / American College of Occupational and Environmental Medicine* 58, no. 5: 436–43.

Lee, Yu-Mi, D. R. Jacobs Jr., and Duk-Hee Lee. 2018. "Persistent Organic Pollutants and Type 2 Diabetes: A Critical Review of Review Articles." *Frontiers in Endocrinology* 27, no. 9: 712.

Leigh, J. Paul, Stephen A. McCurdy, and Marc B. Schenker. 2001. "Costs of Occupational Injuries in Agriculture." *Public Health Reports* 116, no. 3: 235–248.

León-Olea, Martha, Christopher J. Martuniuk, Edward F. Orlando, Mary Ann Ottinger, Cheryl Rosenfeld, Jennifer Wolstenholme, and Vance L. Trudeau. 2014. "Current Concepts in Neuroendocrine Disruption." *General and Comparative Endocrinology* 203: 158–173.

Levin, Sam. 2018. "The Man Who Beat Monsanto: 'They Have to Pay for Not Being Honest.'" *Guardian,* September 26, 2018. https://www.theguardian.com/business/2018/sep/25/monsanto-dewayne-johnson-cancer-verdict.

Liboiron, Max, Manuel Tironi, and Nerea Calvillo. 2018. "Toxic Politics: Acting in a Permanently Polluted World." *Social Studies of Science* 48, no. 3: 331–349.

Lipsky, Michael. 2010. *Street-Level Bureaucracy: Dilemmas of the Individual in Public Services.* New York: Russell Sage Foundation.

London, Jonathan, Amanda Fencl, Sara Watterson, Jennifer Jarin, Alfonso Aranda, Aaron King, Camille Pannu, et al. 2018. *The Struggle for Water Justice in California's San Joaquin Valley: A Focus on Disadvantaged Unincorporated Communities.* Davis: University of California–Davis Center for Regional Change. https://regionalchange.ucdavis.edu/sites/g/files/dgvnsk986/files/inline-files/The%20Struggle%20for%20Water%20Justice%20FULL%20REPORT.pdf.

López, Ann Aurelia. 2007. *The Farmworkers' Journey.* Berkeley: University of California Press.

López, Marcos. 2011. "Places in Production: Nature, Farm Work and Farm Worker Resistance in U.S. and Mexican Strawberry Growing Regions." PhD diss., University of California–Santa Cruz.

Low, Setha M., and Sally Engle Merry. 2010. "Engaged Anthropology: Diversity and Dilemmas." *Current Anthropology* 51, no. 2: S203–S226.

Loza, Mireya. 2016. *Defiant Braceros: How Migrant Workers Fought for Racial, Sexual, and Political Freedom.* Chapel Hill: University of North Carolina Press.

Lynd, Staughton. 2013. *Accompanying: Pathways to Social Change.* Oakland, CA: PM Press.

Lyon, Sarah, and Mark Moberg. 2010. "What's Fair? The Paradox of Seeking Justice through Markets." In *Fair Trade and Social Justice: Global Ethnographies,* edited by Sarah Lyon and Mark Moberg, 1–24. New York: New York University Press.

Macaulay, Luke, and Van Butsic. 2017. "Ownership Characteristics and Crop Selection in California Cropland." *California Agriculture* 71, no. 4: 221–230.

Maeckelbergh, Marianne. 2016. "Whose Ethics? Negotiating Ethics and Responsibility in the Field." In *Impulse to Act: A New Anthropology of Resistance and Social Justice,* edited by Othon Alexandrakis, 211–230. Bloomington: Indiana University Press.

Mahler, Sarah. 1995. *American Dreaming: Immigrant Life on the Margins.* Princeton, NJ: Princeton University Press.

Malloy, Timothy, John Froines, Andrea Hricko, Karla Vasquez, and Mason Gamble. 2018. *Governance on the Ground: Evaluating the Role of County Agricultural Commissioners in Reducing Toxic Pesticide Exposure.* Los Angeles: UCLA School of Law. https://law.ucla.edu/sites/default/files/PDFs/Publications/Emmett%20Institute/_CEN_EMM_PUB_Governance%20on%20the%20Ground.pdf.

Manderson, Lenore, and Carolyn Smith-Morris. 2010. *Chronic Conditions, Fluid States: Chronicity and the Anthropology of Illness.* New Brunswick, NJ: Rutgers University Press.

Manderson, Lenore, and Narelle Warren. 2016. "'Just One Thing after Another': Recursive Cascades and Chronic Conditions." *Medical Anthropology Quarterly* 30, no. 4: 479–497.

Marcus, George E. 1994. "Ethnography in/of the World System: The Emergency of Multisited Ethnography." *Annual Review of Anthropology* 24: 95–117.

Marcus, George E., and Michael M. J. Fischer. 1986. *Anthropology as Cultural Critique: An Experimental Moment in the Human Sciences.* Chicago: University of Chicago Press.

Marks, Amy R., Kim Harley, Asa Bradman, Kathleen Kogut, Dana B. Barr, Caroline Johnson, Norma Calderon, and Brendea Eskenazi. 2010. "Organophosphate Pesticide Exposure and Attention in Young Mexican-American Children." *Environmental Health Perspectives* 118, no. 12: 1768–1774.

Mart, Michelle. 2015. *Pesticides, A Love Story: America's Enduring Embrace of Dangerous Chemicals.* Lawrence: University Press of Kansas.

Martin, Emily. 2013. "Enhancing the Public Impact of Ethnography." *Social Science and Medicine* 99: 205–208.

Martin, Liam. 2015. "Ethnography as a Research Justice Strategy." In *Research Justice: Methodologies for Social Change,* edited by Andrew J. Jolivette, 33–42. Chicago: Policy Press.

Martin, Philip. 2017. "The H2-A Farm Guestworker Program Is Expanding Rapidly." *Working Economics Blog,* Economic Policy Institute, April 13, 2017. https://www.epi.org/blog/h-2a-farm-guestworker-program-expanding-rapidly/.

———. 2011. "The Costs and Benefits of a Raise for Field Workers." *New York Times,* September 30, 2011. https://www.nytimes.com/roomfordebate/2011/08/17/could-farms-survive-without-illegal-labor/the-costs-and-benefits-of-a-raise-for-field-workers.

———. 2007. *Farm Labor Shortages: How Real? What Response.* Washington, DC: Center for Immigration Studies. https://cis.org/sites/cis.org/files/articles/2007/back907.pdf.

Martin, Philip and J. Edward Taylor. 2013. "Ripe with Change: Evolving Farm Labor Markets in the United States, Mexico, and Central America." White paper. Davis, CA: Migration Policy Institute, 2013. https://www.migrationpolicy.org/research/ripe-change-evolving-farm-labor-markets-united-states-mexico-and-central-america

Maskens, Maïté, and Ruy Blanes. 2013. "Don Quixote's Choice: A Manifesto for a Romanticist Anthropology." *HAU: Journal of Ethnographic Theory* 3, no. 3: 245–281.

Mauss, Marcel. (1925) 1967. *The Gift: Forms and Functions of Exchange in Archaic Societies.* Translated by Ian Cunnison. New York: Norton.

Mayer, Brian M., Joan Flocks, and Paul Monaghan. 2010. "The Role of Employers and Supervisors in Promoting Pesticide Safety Behavior among Florida Farmworkers." *American Journal of Industrial Medicine* 53, no. 8: 814–824. https://doi.org/10.1002/ajim.20826.

Mazak, Catherine M. 2017. "Commentary: What Should the University Be in a Time of Devastation?" *Chronicle of Higher Education,* October 26, 2017. https://www.chronicle.com/article/What-Should-the-University-Be/241562.

MCAC (Monterey County Agricultural Commissioner). 2017. *Monterey County Crop Report 2017.* Salinas, CA: MCAC. https://www.co.monterey.ca.us/Home/ShowDocument?id=65737.

McCurdy, Heather H., J. Josh Snodgrass, Charles R. Martinez Jr., Erica C. Squires, Roberto A. Jimenez, Laura E. Isiordia, J. Mark Eddy, Thomas W. McDade, and Jeon Small. 2015. "Stress, Place, and Allostatic Load among Mexican Immigrant Farmworkers in Oregon." *Journal of Immigrant and Minority Health* 17, no. 5: 1518–1525.

McKenna, Brian. 2010. "Exposing Environmental Health Deception as a Government Whistleblower: Turning Critical Ethnography into Public Pedagogy." *Policy Futures in Education* 8, no. 1: 22–36.

———. 2004. "US: Dow's Knowledge Factory." CorpWatch, February 11, 2004. https://corpwatch.org/article/us-dows-knowledge-factory.

McWilliams, Carey. 1996. *Factories in the Field: The Story of Migratory Farm Labor in California.* Berkeley: University of California Press.

Medicine, Beatrice. 2001. *Learning to Be an Anthropologist and Remaining "Native": Selected Writings.* Urbana: University of Illinois Press.

Meghani, Zahra, and Jennifer Kuzma. 2011. "The 'Revolving Door' between Regulatory Agencies and Industry: A Problem That Requires Reconceptualizing Objectivity." *Journal of Agricultural and Environmental Ethics* 24, no. 6: 575–599.

Mehrpour, Omid, Parissa Karrari, Nasim Zamani, Aristedes M. Tsatsakis, and Mohammad Abdollahi. 2014. "Occupational Exposure to Pesticides and Consequences on Male Semen and Fertility: A Review." *Toxicology Letters* 230, no. 2: 146–156.

Mendenhall, Emily. 2012. *Syndemic Suffering: Social Distress, Depression, and Diabetes among Mexican Immigrant Women.* New York: Routledge.

Mie, Axel, Christina Rudén, and Philippe Grandjean. 2018. "Safety of Safety Evaluation of Pesticides: Developmental Neurotoxicity of Chlorpyrifos and Chlorpyrifos-Methyl." *Environmental Health* 17, no. 77: 1–5.

Migrant Clinicians Network. 2017. "Protecting Pregnant Agricultural Workers: Medical-Legal Partnership in California." Accessed September 7, 2020. https://www.migrantclinician.org/streamline/2018/protecting-pregnant-agricultural-workers-medical-legal-partnership.

Miner, Horace. 1956. "Body Rituals among the Nacirema." *American Anthropologist* 58, no. 3: 503–507.

Minkoff-Zern, Laura-Anne. 2019. *The New American Farmer: Immigration, Race, and the Struggle for Sustainability.* Cambridge, MA: MIT Press, 2019.

Mitchell, Don. 1996. *The Lie of the Land: Migrant Workers and the California Landscape.* Minneapolis: University of Minnesota Press.

Mnif, Wiseem, Aziza Ibn Hadj Hassine, Aicha Bouaziz, Aghleb Bartegi, Oliver Thomas, and Benoit Roig. 2011. "Effect of Endocrine Disruptor Pesticides: A Review." *International Journal of Environmental Research in Public Health* 8, no. 6: 2265–2203.

Mohan, Geoffrey. 2018. "How California's Farm Labor Shortage Made Friends of Old Rivals." *Los Angeles Times*, July 6, 2018. https://www.latimes.com/business/la-fi-ufw-darrigo-20180706-story.html.

———. 2017. "Dole, the World's Largest Fresh Fruit and Vegetable Company, Is Stepping Back from Southland." *Los Angeles Times*, August 30, 2017. https://www.latimes.com/business/la-fi-dole-socal-20170830-story.html.

Molina, Natalia. 2006. *Fit to Be Citizens? Public Health and Race in Los Angeles, 1879–1939.* Berkeley: University of California Press.

Monaghan, Paul. 2011. "Lessons Learned from a Community Coalition with Diverse Stakeholders: The Partnership for Citrus Worker Health." *Annals of Anthropological Practice* 35, no. 2: 27–42.

Montoya, Michael. 2011. *Making the Mexican Diabetic: Race, Science, and the Genetics of Inequality.* Berkeley: University of California Press.

Morales, Maricela, and Christy Lubin. 2019. "Disaster Relief Should Include All Victims." *Capitol Weekly* (Sacramento, CA), January 10, 2019. https://capitolweekly.net/disaster-relief-all-victims/.

Morehouse, Lisa. 2016. "How a Farmworker 'Company Town' Is Taking Shape in the Salinas Valley." KQED, November 5, 2016. https://www.kqed.org/news/11155240/how-a-farmworker-company-town-is-taking-shape-in-the-salinas-valley.

Morello-Frosch, Rachel, Julia Green Brody, Phil Brown, Rebecca Gasior Altman, Ruthann A. Rudel, and Carla Perez. 2009. "Toxic Ignorance and Right-to-Know in Biomonitoring Results Communication: A Survey of Scientists and Study Participants." *Environmental Health* 8, article 6. http://www.ehjournal.net/content/8/1/6.

Moretto, Angelo, and Claudio Colosio. 2011. "Biochemical and Toxicological Evidence of Neurological Effects of Pesticides: The Example of Parkinson's Disease." *Neurotoxicology* 32, no. 4: 383–91.

Morgan, A., C. M. Baker, J.S.F. Chu, K. Lee, B. A. Crandall, and L. Jose. 2002. "Production of Herbicide Tolerant Strawberry through Genetic Engineering." In *Acta Horticulturae*, vol. 567, *Proceedings of the Fourth International Strawberry Symposium*, edited by T. Hietaranta, M.-M. Linna, P. Palonen, and P. Parikka, 113–115. Leuven, Belgium: International Society for Horticultural Science.

Moses, Marion. 1999. "Pesticide-Related Health Problems and Farmworkers." *American Association of Occupational Health Nurses Journal* 37, no. 3: 115–130.

Mulligan, Jessica, and Heide Castañeda. 2018. *Unequal Coverage: The Experience of Health Care Reform in the United States*. New York: New York University Press.

Murphy, Hugh. n.d. "Food Indigenous to the Western Hemisphere—Strawberry." American Indian Health and Diet Project. Accessed September 7, 2020. http://www.aihd.ku.edu /foods/strawberry.html.

Murphy, Michelle. 2008. "Chemical Regimes of Living." *Environmental History* 13, no. 4: 695–703.

———. 2006. *Sick Building Syndrome and the Problem of Uncertainty: Environmental Politics, Technoscience, and Women Workers*. Durham, NC: Duke University Press.

Myerhoff, Barbara. 2007. *Stories as Equipment for Living: Last Talks and Tales of Barbara Myerhoff*. Edited by Marc Kaminsky and Mark Weisse. Ann Arbor: University of Michigan Press.

Myers, Justin Sean, and Joshua Sbicca. 2015. "Bridging Good Food and Good Jobs: From Succession to Confrontation within Alternative Food Movements." *Geoforum* 61: 17–26.

Myers, Robert. 2011. "The Familiar Strange and the Strange Familiar in Anthropology and Beyond." *General Anthropology* 18, no. 2: 1–9.

Nader, Laura. 2017. "Unravelling the Politics of Silencing." *Public Anthropologist Journal Blog*, October 11, 2017. http://publicanthropologist.cmi.no/2017/10/11/academic-politics-of -silencing/#laura-nader.

———. 2002. *The Life of the Law: Anthropological Projects*. Berkeley: University of California Press.

———. 1999. "Thinking Public Interest Anthropology 1890s–1990s." *General Anthropology* 5, no. 2: 1, 7–9.

———. 1997. "Controlling Processes: Tracing the Dynamic Components of Power." *Current Anthropology* 38, no. 5: 711–737.

———. 1972. "Up the Anthropologist: Perspectives Gained from Studying Up." In *Reinventing Anthropology*, edited by Dell Hymes, 285–311. New York: Pantheon.

Nash, June. 1993. *We Eat the Mines and the Mines Eat Us: Dependency and Exploitation in the Bolivian Tin Mines*. New York: Columbia University Press.

NCFH (National Center for Farmworker Health). 2018. *Indigenous Agricultural Workers*. Buda, TX: NCFH. http://www.ncfh.org/uploads/3/8/6/8/38685499/fs-indigenous_ag _workers_2018.pdf.

———. 2012. *Farmworker Health Factsheet: Demographics*. Buda, TX: NCFH. http://www .ncfh.org/uploads/3/8/6/8/38685499/fs-migrant_demographics.pdf.

Nestle, Marion. 2018. *Unsavory Truth: How Food Companies Skew The Science of What We Eat*. New York: Basic Books.

———. 2013. *Food Politics: How the Food Industry Influences Nutrition and Health*. Berkeley: University of California Press.

Neuburger, Bruce. 2013. *Lettuce Wars: Ten Years of Work and Struggle in the Fields of California.* New York: Monthly Review Press.

Newhall, A. G. 1955. "Disinfestation of Soil by Heat, Flooding and Fumigation." *Botanical Review* 21, no. 4: 189–250.

Ngai, Mae M. 2014. *Impossible Subjects: Illegal Aliens and the Making of Modern America.* Princeton, NJ: Princeton University Press.

Nguyen, Vinh-Kim. 2010. *The Republic of Therapy: Triage and Sovereignty in West Africa's Time of AIDS.* Durham, NC: Duke University Press.

Nicolopoulou-Stamati, Polyxeni, Sotirios Maipas, Chrysanthi Kotampasi, Panagiotis Stamatis, and Luc Hens. 2016. "Chemical Pesticides and Human Health: The Urgent Need for a New Concept in Agriculture." *Frontiers in Public Health* 4, no. 148: 1–8.

Nixon, Rob. 2011. *Slow Violence and the Environmentalism of the Poor.* Cambridge, MA: Harvard University Press.

OEHHA (California Office of Environmental Health Hazard Assessment). 2017. *Medical Supervision of Pesticide Workers: Guidelines for Physicians Who Supervise Workers Exposed to Cholinesterase Inhibiting Pesticides.* Edition 6.0. Sacramento: Pesticide and Environmental Toxicology Branch, OEHHA. https://oehha.ca.gov/media/downloads/pesticides/document-pesticides/physicianguidelines.pdf.

———. n.d.-a. "About Proposition 65." Web page, accessed September 7, 2020. https://oehha.ca.gov/proposition-65/about-proposition-65.

———. n.d.-b. "Pesticide Use." Web page, accessed September 7, 2020. https://oehha.ca.gov/calenviroscreen/indicator/pesticide-use.

Olson, Valerie. 2018. *Into the Extreme: U.S. Environmental Systems and Politics beyond Earth.* Minneapolis: University of Minnesota Press.

———. 2010. "The Ecobiopolitics of Space Biomedicine." *Medical Anthropology* 29, no. 2: 170–193.

Önel, Gülcan, and Derek Farnsworth. 2016. "Guest Workers: Past, Present and Future." *Citrus Industry,* May 2016, 12–13.

Oreskes, Naomi, and Erik M. Conway. 2010. *Merchants of Doubt: How a Handful of Scientists Obscured the Truth on Issues from Tobacco Smoke to Global Warming.* New York: Bloomsbury.

Orñelas, Jesse. 2019. "The Heart of Our Work Should Be Transformation." Youth Leadership Institute, September 25, 2019. https://yli.org/2019/09/the-heart-of-our-work-should-be-transformation/?fbclid=IwAR0JwAIk2adM8HtHykzjlOAdydBiGzgZtonqSpftG9oStbDjdO-CofvN4a8.

Otero, Gerardo. 2018. *The Neoliberal Diet: Healthy Profits, Unhealthy People.* Austin: University of Texas Press.

———. 2011. "Neoliberal Globalization, NAFTA and Migration: Mexico's Loss of Food and Labor Sovereignty." *Journal of Poverty* 15, no. 4: 384–402.

Oxfam America. n.d. "Food Security, Agriculture and Livelihoods." Accessed September 7, 2020. https://policy-practice.oxfamamerica.org/work/food-agriculture-livelihoods/.

Oxfam International. 2018. *Ripe for Change: Ending Human Suffering in Supermarket Supply Chains.* Oxford: Oxfam International. https://d1tn3vj7xz9fdh.cloudfront.net/s3fs-public/file_attachments/cr-ripe-for-change-supermarket-supply-chains-210618-en.pdf.

Padula, Amy M., Hongtai Huang, Rebecca J. Baer, Laura M. August, Marta M. Jankowska, Laura L. Jeliffe-Pawlowski, Mariana Sirota, and Tracey J. Woodruff. 2018. "Environmental Pollution and Social Factors as Contributors to Preterm Birth in Fresno County." *Environmental Health* 17, no. 1: 70.

Park, Sun Hee, and David N. Pellow. 2011. *The Slums of Aspen: Immigrants vs. the Environment in America's Eden.* New York: New York University Press, 2011.

Passel, Jeffrey S., D'Vera Cohn, and Ana Gonzalez-Barrera. 2012. "Net Migration from Mexico Falls to Zero—And Perhaps Less." Pew Research Center, April 23, 2012. https://www.pewresearch.org/hispanic/2012/04/23/net-migration-from-mexico-falls-to-zero-and-perhaps-less/

Patel, Raj, and Jason Moore. 2017. *A History of the World in Seven Cheap Things: A Guide to Capitalism, Nature, and the Future of the Planet.* Berkeley: University of California Press, 2017.

Pavlovitz, John. 2019. "The Privilege of Positivity." *Stuff That Needs to Be Said* (blog), July 6, 2019. https://johnpavlovitz.com/2019/07/06/the-privilege-of-positivity/?fbclid=IwAR2Uzwl16snteAOt13vRcOuc5eKu_YvNHccu-R3JxR4Fz-FT3CTfA_l132k.

Pels, Peter. 2014. "After Objectivity: An Historical Approach to the Intersubjective in Ethnography." *HAU: Journal of Ethnographic Theory* 4, no. 1: 211–236.

Peña, Devon G. 2011. "Structural Violence, Historical Trauma, and Public Health: The Environmental Justice Critique of Contemporary Risk Science and Practice." In *Communities, Neighborhoods and Health: Expanding the Boundaries of Place*, edited by Linda M. Burton, Susan P. Kemp, ManChui Leung, Stephen A. Matthews, and David T. Takeuchi, 203–218. New York: Springer.

———. 2005. *Mexican Americans and the Environment: Tierra y Vida.* Tuscon: University of Arizona Press.

Pesticide Action Network. n.d.-a. "Fenpyroximate." Web page, accessed September 7, 2020. http://www.pesticideinfo.org/Detail_Chemical.jsp?Rec_Id=PRI3213.

———. n.d.-b. "Fumigants." Web page, accessed August 31, 2012. https://web.archive.org/web/20120922093032/http://www.panna.org/resources/specific-pesticides/fumigants.

———. n.d.-c. "Pesticide Action Network (PAN) Bad Actor Pesticides." Web page, accessed September 7, 2020. http://www.pesticideinfo.org/Docs/ref_toxicity7.html#BadActor.

Pezzoli, Gianni, and Emanuele Cereda. 2013. "Exposure to Pesticides or Solvents and Risk of Parkinson Disease." *Neurology* 80, no. 22: 2035–2041.

Phillips, James. 2010. "Resource Access, Environmental Struggles and Human Rights in Honduras." In *Life and Death Matters*, 2nd ed., edited by Barbara Rose Johnston, 209–232. Walnut Creek, CA: Left Coast.

Philpott, Tom. 2011. "Is the 'Clean 15' Just as Toxic as the 'Dirty Dozen?'" *Mother Jones*, June 21, 2011. https://www.motherjones.com/food/2011/06/update-dirty-dozen-pesticides-and-farm-workers/.

———. 2007. "An EPA-Approved Pesticide Is Worse Than the One It's Replacing." *Grist*, December 7, 2007. https://grist.org/article/sterile-soil-dirty-hands/.

Piazza, Mauri José, and Almir Antônio Urbanetz. 2019. "Environmental Toxins and the Impact of Other Endocrine Disrupting Chemicals in Women's Reproductive Health." *JBRA Assisted Reproduction* 23, no. 2: 154–164.

Pine, Adrienne. 2013. "Revolution as Care Plan: Ethnography, Nursing and Somatic Solidarity in Honduras." *Social Science and Medicine* 99: 143–152.

Pitkin Derose, Kathryn, José J. Escarce, and Nicole Lurie. 2007. "Immigrants and Health Care: Sources of Vulnerability." *Health Affairs* 26, no. 5: 1258–1268.

Poison Papers. n.d. "The Poison Papers Project." Web page, accessed September 7, 2020. https://www.poisonpapers.org/about-us/.

Pollan, Michael. 2001. "Naturally." *New York Times Magazine*, May 13, 2001. https://michaelpollan.com/articles-archive/naturally/.

Powdermaker, Hortense. 1966. *Stranger and Friend: The Way of an Anthropologist.* New York: W. W. Norton.

Prasad, Rose Schmitz, Andrew Slocombe, Robbie Welling, Walker Wieland, and Lauren Zeise. 2017. *CalEnviroScreen 3.0: Update to the California Communities Environmental Health Screening*

Tool. Sacramento, CA: California Environmental Protection Agency and Office of Environmental Health Hazard Assessment.
https://oehha.ca.gov/media/downloads/calenviroscreen/report/ces3report.pdf.

Price, David H. 2004. *Threatening Anthropology: McCarthyism and the FBI's Surveillance of Activist Anthropologists*. Durham, NC: Duke University Press.

Pulido, Laura. 2017. "Geographies of Race and Ethnicity II: Environmental Racism, Racial Capitalism and State-Sanctioned Violence." *Progress in Human Geography* 41, no. 4: 524–533.

———. 1996. *Environmentalism and Economic Justice: Two Chicano Struggles in the Southwest*. Tucson: University of Arizona Press, 1996.

Pundir, Pallavi, and Rohit Jain. 2019. "Powerful Photos of the Generation Born after the Bhopal Gas Tragedy," *Vice*, November 29, 2019. https://www.vice.com/en_in/article/zmjp34/powerful-photos-of-the-generation-born-after-the-bhopal-gas-tragedy.

Quackenbush, Robin, Barbara Hackley, and Jane Dixon. 2006. "Screening for Pesticide Exposure: A Case Study." *Journal of Midwifery and Women's Health* 51, no. 1: 3–11.

Quandt, Sara A., Thomas A. Arcury, Colin K. Austin, and Luis F. Cabrera. 2001. "Preventing Occupational Exposure to Pesticides: Using Participatory Research with Latino Farmworkers to Develop an Intervention." *Journal of Immigrant Health* 3, no. 2: 85–96.

Quandt, Sara A., Thomas A. Arcury, Colin K. Austin, and Rosa M. Saavedra. 2000. "Farmworker and Farmer Perceptions of Farmworker Agricultural Chemical Exposure in North Carolina." In *Illness and the Environment: A Reader in Contested Medicine*, edited by Steve Kroll-Smith, 175–192. New York: New York University Press.

Quandt, Sara A., Francis O. Walker, Jennifer W. Talton, Haiying Chen, and Thomas A. Arcury. 2017. "Olfactory Function in Latino Farmworkers over Two Years: Longitudinal Exploration of Subclinical Neurological Effects of Pesticide Exposure." *Journal of Occupational and Environmental Medicine* 59, no. 12: 1148–1152.

Quesada, James. 2011. "No Soy Welferero: Undocumented Latino Laborers in the Crosshairs of Legitimation Maneuvers." *Medical Anthropology* 30, no. 4: 386–408.

Quesada, James, Laurie Kain Hart, and Philippe Bourgois. 2011. "Structural Vulnerability and Health: Latino Migrant Laborers in the United States." *Medical Anthropology* 30, no. 4: 339–362.

Quintero-Somaini, Adrianna, and Mayra Quirindongo. 2004. "Hidden Danger: Environmental Health Threats in the Latino Community." Washington, DC: National Resources Defense Council, 2004. https://www.nrdc.org/sites/default/files/latino_en.pdf.

Quirós-Alcalá, Lesliam, Asa Bradman, Marcia Nishioka, Martha E. Harnly, Alan Hubbard, Thomas E. McKone, Jeannette Ferber, and Brenda Eskenazi. 2011. "Pesticides in House Dust from Urban and Farmworker Households in California: An Observational Measurement Study." *Environmental Health* 10, no. 1: 1–15.

Raanan, Rachel, John R. Balmes, Kim G. Harley, Robert B. Gunier, Sheryl Magzamen, Asa Bradman, and Brenda Eskenazi. 2016. "Decreased Lung Function in 7-Year-Old Children with Early-Life Organophosphate Exposure." *Thorax* 71, no. 2: 148–153.

Radio Bilingüe. 2015. "Sindicalización de Jornaleros en San Quintín (Primera)." Radio Bilingüe, November 23, 2015. http://radiobilingue.org/features/sindicalizacion-de-jornaleros-en-san-quintin-primera/.

Raine, George. 2008. "China Gets a Taste of California Strawberries." *SF Gate*, August 17, 2008. https://www.sfgate.com/business/article/China-gets-a-taste-of-California-strawberries-3273022.php.

Rajak, Dinah. 2010. "'HIV/AIDS Is Our Business': The Moral Economy of Treatment in a Transnational Mining Company." *Journal of the Royal Anthropological Institute* 16, no. 3: 551–71.

Ramey, Madeline. 2019. "Farmworkers' Low Wage Rates Have Risen Modestly; Now Congress May Pass a Law to Lower Them." *Harvesting Justice* (blog), August 23, 2019. https://www.farmworkerjustice.org/blog-post/farmworkers-low-wage-rates-have-risen-modestly-now-congress-may-pass-a-law-to-lower-them/.

Rao, Pamela, Thomas A. Arcury, Sara A. Quandt, and Alicia Doran. 2004. "North Carolina Growers' and Extension Agents' Perceptions of Latino Farmworker Pesticide Exposure." *Human Organization* 63, no. 2: 151–161.

Rao, Pamela, Sara A. Quandt, Alicia M. Doran, Beverly M. Snively, and Thomas A. Arcury. 2007. "Pesticides in the Homes of Farmworkers: Latino Mothers' Perceptions of Risk to Their Children's Health." *Health Education and Behavior* 34, no. 2: 335–353.

Rappazzo, Kristen M., Joshua L. Warren, Robert E. Meyer, Amy H. Herring, Alison P. Sanders, Naomi C. Brownstein, and Thomas J. Luben. 2016. "Maternal Residential Exposure to Agricultural Pesticides and Birth Defects in a 2003 to 2005 North Carolina Birth Cohort." *Clinical and Molecular Teratology* 106, no. 4: 240–249.

Reese, Ashanté. 2019. "When We Come to Anthropology, Elsewhere Comes with Us." Anthropology News, February 20, 2019. http://www.anthropology-news.org/index.php/2019/02/20/elsewhere-comes-with-us/.

Remy, Anselme. 1976. "Anthropology: For Whom and What?" *Black Scholar: Journal of Black Studies and Research* 7, no. 7: 12–16.

Richter, Lauren. 2017. "Constructing Insignificance: Critical Race Perspectives on Institutional Failure in Environmental Justice Communities." *Environmental Sociology* 4, no. 1: 107–121.

Rivera-Salgado, Gaspar. 2014. "Transnational Indigenous Communities: The Intellectual Legacy of Michael Kearney." *Latin American Perspectives* 41, no. 3: 26–46.

Robb, Erika L., and Mari B. Baker. 2019. "Organophosphate Toxicity." In *StatPearls*, online ed. Treasure Island, FL: StatPearls. https://www.ncbi.nlm.nih.gov/books/NBK470430/.

Robertson-Steel, Iain. 2006. "Evolution of Triage Systems." *Emergency Medical Journal* 23, no. 2: 154–55.

Robinson, Jo. 2013. *Eating on the Wild Side: The Missing Link to Optimum Health*. New York: Little, Brown.

Rodriguez-Delgado, Cresencio. 2019a. "Chemical Exposures in California's Vast Cropland Spark Fear for Growers and Workers." *Fresno (CA) Bee*, September 20, 2019. https://www.fresnobee.com/news/local/article234705742.html.

———. 2019b. "Dozens of Farm Workers Exposed to Pesticides in South Valley. 3 Sent to Hospital." *Fresno (CA) Bee*, June 18, 2019. https://www.fresnobee.com/news/local/article231701243.html.

Rohrlich, Justin. 2019. "This Is How Much It Costs to Be Smuggled over the US Border." Quartz, June 10, 2019. https://qz.com/1632508/this-is-how-much-it-costs-to-cross-the-us-mexico-border-illegally/.

Rosas, Lisa G., and Brenda Eskenazi. 2008. "Pesticides and Child Neurodevelopment." *Current Opinions in Pediatrics* 20, no. 2: 191–197.

Ruiz, Daniel, Marisol Becerra, Jyotsna S. Jagai, Kerry Ard, and Robert M. Sargis. 2018. "Disparities in Environmental Exposures to Endocrine-Disrupting Chemicals and Diabetes Risk in Vulnerable Populations." *Diabetes Care* 41, no. 1: 193–205.

Runkle, Jennifer D., J. Antonio Tovar-Aguilar, Eugenia Economos, Joan Flocks, Bryan Williams, Juan F. Muniz, Marie Semple, and Linda McCauley. 2013. "Pesticide Risk Perception and Biomarkers of Exposure in Florida Female Farmworkers." *Journal of Occupational and Environmental Medicine* 55, no. 11: 1286–1292.

Rusnak, Paul. 2018. "Arysta LifeScience to Be Sold in Billion-Dollar Deal." *Growing Produce*, July 20, 2018. https://www.growingproduce.com/citrus/arysta-lifescience-to-be-sold-in -billion-dollar-deal/.

Rylko-Bauer, Barbara, and Paul Farmer. 2002. "Managed Care or Managed Inequality? A Call for Critiques of Market-Based Medicine." *Medical Anthropology Quarterly* 16, no. 4: 476–502.

Salomón J., Amrah. 2015. "Telling to Reclaim, Not to Sell: Resistance Narratives and the Marketing of Justice." In *Research Justice: Methodologies for Social Change*, edited by Andrew J. Jolivette, 185–198. Chicago: Policy Press.

Salvatore, Alicia L., Asa Bradman, Rosemary Castorina, José Camacho, Jesús López, Dana B. Barr, John Snyder, Nicholas P. Jewell, and Brenda Eskenazi. 2008. "Occupational Behaviors and Farmworkers' Pesticide Exposure: Findings from a Study in Monterey County, California." *American Journal of Industrial Medicine* 51, no. 10: 782–794.

Salvatore, Alicia L., Rosemary Castorina, José Camacho, Norma Morga, Jesús López, Marcia Nishioka, Dana B. Barr, Brenda Eskenazi, and Asa Bradman. 2015. "Home-Based Community Health Worker Intervention to Reduce Pesticide Exposures to Farmworkers' Children: A Randomized Controlled Trial." *Journal of Exposure Science and Environmental Epidemiology* 25, no. 6: 608–15.

Salvatore, Alicia L., Johnathan Chevrier, Asa Bradman, José Camacho, Jesús López, Geri Kavanagh-Baird, Meredith Minkler, and Brenda Eskenazi. 2009. "A Community-Based Participatory Worksite Intervention to Reduce Pesticide Exposures to Farmworkers and Their Families." *American Journal of Public Health* 99, no. 3: S578–S581.

Sanborn, Margaret, Kathleen J. Kerr, Luz Helena Sanin, Donald C. Cole, Kate L. Bassil, and Cathy Vakil. 2007. "Non-cancer Health Effects of Pesticides: Systematic Review and Implications for Family Doctors." *Canadian Family Physician* 53, no. 10: 1712–1720.

Sanchez, Irene. 2019. "El Paso Horror Spotlights Long History of Anti-Latino Violence in the US." CNN, August 5, 2019. https://www.cnn.com/2019/08/05/opinions/el-paso -shooting-was-a-hate-crime-against-latinos-sanchez/index.html.

Sanford, Victoria. 2006. "Introduction." In *Engaged Observer: Anthropology, Advocacy, and Activism*, edited by Victoria Sanford and Asale Angel-Ajani, 1–15. New Brunswick, NJ: Rutgers University Press.

Sangaramoorthy, Thurka. 2019. "Liminal Living: Everyday Injury, Disability and Instability among Migrant Mexican Women in Maryland's Seafood Industry." *Medical Anthropology Quarterly* 33, no. 4: 557–578.

Sangari, Kumkum. 1986. "The Politics of the Possible." *Cultural Critique* 7: 157–186.

Sanjek, Roger. 2015. "Mutuality and Anthropology: Terms and Modes of Engagement." In *Mutuality: Anthropology's Changing Terms of Engagement*, edited by Roger Sanjek, 285–310. Philadelphia: University of Pennsylvania Press.

Saxton, Dvera I. 2015a. "Beyond the Fork: If We Are What We Eat, How Can We Continue to Stomach Farmworker Injustice?" *Blue Review* (Boise State University), October 5, 2015. https://wayback.archive-it.org/8092/20190724153750/https://thebluereview.org/strawberry -harvest-and-farmworker-injustice/.

———. 2015b. "Ethnographic Movement Methods: Anthropology Takes on the Pesticide Industry." *Journal of Political Ecology* 22, no. 1: 368–388.

———. 2015c. "Strawberry Fields as Extreme Environments: The Ecobiopolitics of Farmworker Health." *Medical Anthropology* 34, no. 2: 166–183.

———. 2013. "Layered Disparities, Layered Vulnerabilities: Farmworker Health and Agricultural Corporate Power on and off the Farm." PhD diss., American University, 2013.

———. 2012. "Op-Ed: Stop and Sniff the Marigolds and Brassica: Healthy Communities Are Essential for Viable Economies, Not Methyl Iodide." *Californian* (Salinas, CA), March 27, 2012.

———. 2011. "Op-Ed: Methyl Iodide: Coming to a Field Near You?" *Register-Pajaronian* (Watsonville, CA), November 5, 2011.

Saxton, Dvera I., Phil Brown, Samaras Seguinot-Medina, Lorraine Eckstein, David O. Carpenter, Pamela Miller, and Vi Waghiyi. 2015. "Commentary: Environmental Health and Justice and the Right to Research: Institutional Review Board Denials of Community-Based Chemical Biomonitoring of Breast Milk." *Environmental Health* 14, article 90. https://ehjournal.biomedcentral.com/articles/10.1186/s12940-015-0076-x.

Saxton, Dvera, and Victoria Sanchez. 2016. "Ground Truthing the Central Valley: Introduction to the Series on Student Environmental Ethnographic Journalism." *Engagement* (blog), August 9, 2016. https://aesengagement.wordpress.com/2016/08/09/ground-truthing-the-central-valley-introduction-to-the-series-on-student-environmental-ethnographic-journalism/.

Saxton, Dvera, I., and Angela Stuesse. 2018. "Workers' Decompensation: Engaged Research with Injured Im/Migrant Workers." *Anthropology of Work Review* 39, no. 2: 65–78.

Scammell, Madeleine Kangsen, Laura Senier, Jennifer Darrah-Okike, Phil Brown, and Susan Santos. 2009. "Tangible Evidence, Trust and Power: Public Perceptions of Community Environmental Health Studies." *Social Science and Medicine* 68, no. 1: 145–153.

Schelhas, John. 2002. "Race, Ethnicity and Natural Resources in the United States: A Review." *Natural Resources Journal* 42, no. 4: 723–763.

Scheper-Hughes, Nancy. 1995. "The Primacy of the Ethical: Propositions for a Militant Anthropology." *Current Anthropology* 36, no. 3: 409–440.

———. 1992. *Death without Weeping: The Violence of Everyday Life in Brazil*. Berkeley: University of California Press.

———. 1990. "Three Propositions for a Critically Applied Medical Anthropology." *Social Science and Medicine* 30, no. 2: 189–197.

Schlosser, Eric. 2003. "In the Strawberry Fields." In *Reefer Madness: Sex, Drugs, and Cheap Labor in the American Black Market*, 77–108. New York: Houghton Mifflin.

———. 2001. *Fast Food Nation: The Dark Side of the All-American Meal*. New York: Houghton Mifflin.

Schuller, Mark. 2010. "From Activist to Applied Anthropologist to Anthropologist? On the Politics of Collaboration." *Practicing Anthropology* 32, no. 1: 43–47.

Scott, James. 1998. *Seeing Like a State: How Certain Schemes to Improve the World Have Failed*. New Haven, CT: Yale University Press, 1998.

Seabrook, John. 2019. "The Age of Robot Farmers." *New Yorker*, April 8, 2019. https://www.newyorker.com/magazine/2019/04/15/the-age-of-robot-farmers?utm_campaign=aud-dev&utm_source=nl&utm_brand=tny&utm_mailing=TNY_Magazine_Daily_040819&utm_medium=email&bxid=5be9e0743f92a40469f1a62e&user_id=53538562&esrc=&utm_term=TNY_Daily.

Sen, Debarati. 2017. *Everyday Sustainability: Gender Justice and Fair Trade Tea in Darjeeling*. Albany: State University of New York Press.

Seufert, Verena, Anavin Ramankutty, and Jonathan A. Foley. 2012. "Comparing the Yields of Organic and Conventional Agriculture." *Nature* 485: 229–232.

Shaw, Susan. 2009. *Governing How We Care: Contesting Community and Defining Differences in U.S. Public Health Programs*. Philadelphia: Temple University Press.

Sherman, Jocelyn. 2018. "Facing Labor Shortages and Inaction on Immigration in D.C., UFW & D'Arrigo Take Novel Local Approach with Breakthrough Union Contract." United Farm Workers, June 28, 2018. https://ufw.org/darrigocontract/.

Silverman, Sydel. 2007. "American Anthropology in the Middle Decades: A View from Hollywood." *American Anthropologist* 109, no. 3: 519–528.

Singer, Merrill. 2011. "Down Cancer Alley: The Lived Experience of Health and Environmental Suffering in Louisiana's Chemical Corridor." *Medical Anthropology Quarterly* 25, no. 2: 141–163.

———. 1995. "Beyond the Ivory Tower: Critical Praxis in Medical Anthropology." *Medical Anthropology Quarterly* 9, no. 1: 80–106.

———. 1986. "The Emergence of a Critical Medical Anthropology." *Medical Anthropology* 17, no. 5: 128–129.

Singer, Merrill, and Arachu Castro. 2004. "Anthropology and Health Policy: A Critical Perspective." In *Unhealthy Health Policy: A Critical Anthropological Examination*, edited by Arachu Castro and Merrill Singer, xi–xx. Walnut Creek, CA: Altamira.

Smith-Nonini, Sandy. 2011. "The Illegal and the Dead: Are Mexicans Renewable Energy?" *Medical Anthropology* 30, no. 5: 454–474.

———. 2010. *Healing the Body Politic: El Salvador's Popular Struggle for Health Rights from Civil War to Neoliberal Peace*. New Brunswick, NJ: Rutgers University Press.

Snipes, Shedra A., Sharon P. Cooper, and Eva M. Shipp. 2017. "'The Only Thing I Wish I Could Change Is That They Treat Us like People and Not Like Animals': Injury and Discrimination among Latino Farmworkers." *Journal of Agromedicine* 22, no. 1: 36–46.

Snipes, Shedra Amy, Beti Thompson, Kathleen O'Connor, Bettina Shell-Duncan, Denae King, Angelica P. Herrera, and Bridgette Navarro. 2009. "'Pesticides Protect the Fruit, but Not the People': Using Community-Based Ethnography to Understand Farmworker Pesticide-Exposure Risks." *American Journal of Public Health* 99, no. 3: S616–S621.

Solomon, Harris. 2017. "Shifting Gears: Triage and Traffic in Urban India." *Medical Anthropology Quarterly* 31, no. 3: 349–364.

Stack, Carol B. 1970. *All Our Kin: Strategies for Survival in a Black Community*. New York: Harper and Row.

Steinberg, Sheila Lakshmi, and Steven J. Steinberg. 2008. *People, Place and Health: A Sociospatial Perspective of Agricultural Workers and Their Environment*. Arcata: California Center for Rural Policy, Humboldt State University.

Steinberg, Steven J., and Sheila Lakshmi Steinberg. 2008. *People, Place and Health: A Pesticide Atlas of Monterey and Tulare County, California*. Arcata: California Center for Rural Policy, Humboldt State University.

Steinhauer, Jennifer. 2008. "An Olympic Feat for Strawberries." *New York Times*, August 6, 2008. https://www.nytimes.com/2008/08/06/sports/olympics/06berry.html.

Stephen, Lynn. 2017. "Bearing Witness: Testimony in Latin American Anthropology." *Journal of Latin American and Caribbean Anthropology* 22, no. 1: 85–109.

———. 2007. *Transborder Lives: Indigenous Oaxacans in Mexico, California, and Oregon*. Durham, NC: Duke University Press.

Stewart, Jessica. 2019. "15 Facts You Need to Know about the Delightfully Weird 'Garden of Earthly Delights.'" *My Modern Met* (blog), July 6, 2019. https://mymodernmet.com /garden-of-earthly-delights-bosch/.

Stoicheff, Lily. 2018. "The Impact of Santa Cruz County Agriculture's Labor Shortage." *Good Times* (Santa Cruz, CA), April 3, 2018. http://goodtimes.sc/cover-stories/impact-santa -cruz-county-agriculture-labor-shortage/.

Stoll, Steven. 1994. *The Fruits of Natural Advantage: Making the Industrial Countryside in California*. Berkeley: University of California Press.

Straut-Eppsteiner, Holly. 2020. "Trump's Public Charge Rule Created Harm Even Before It Was Implemented." Blog, *National Immigration Law Center*, March 2, 2020. https://www .nilc.org/2020/03/02/public-charge-rule-created-harm-before-it-was-implemented/

Street, Richard Steven. 2004. *Beasts of the Field: A Narrative History of California Farmworkers, 1769–1913*. Stanford, CA: Stanford University Press.

Strolich, Ron, Cathy Wirth, Ana Fernandez Besada, and Christy Getz. 2008. *Farm Labor Conditions on Organic Farms in California*. Davis: California Institute for Rural Studies.

Stuesse, Angela. 2018. "When They're Done with You: Legal Violence and Structural Vulnerability among Injured Immigrant Poultry Workers." *Anthropology of Work Review* 39, no. 2: 79–93.

———. 2016. *Scratching Out a Living: Latinos, Race and Work in the Deep South*. Berkeley: University of California Press.

Stuesse, Angela, Mat Coleman, and Sarah Horton. 2016. "Driving While Latino." *Huffington Post*, September 30, 2016. https://www.huffpost.com/entry/driving-while-latino_b_57ed6ce4e4b07f20daa1052f.

Superstorm Research Lab. 2013. "A Tale of Two Sandys." White paper. New York: Superstorm Research Lab, 2013. https://superstormresearchlab.files.wordpress.com/2013/10/srl-a-tale-of-two-sandys.pdf.

Swartz, Alison, Susan Levine, Hanna-Andrea Rother, and Fritha Langerman. 2018. "Toxic Layering through Three Disciplinary Lenses: Childhood Poisoning and Street Pesticide Use in Cape Town, South Africa." *Medical Humanities* 44, no. 4: 247–252.

Szasz, Andrew. 2007. *Shopping Our Way to Safety: How We Changed from Protecting the Environment to Protecting Ourselves*. Minneapolis: University of Minnesota Press.

Taylor, A. L. 1951. "Chemical Treatment of the Soil for Nematode Control." *Advances in Agronomy* 3: 243–264.

Taylor, Rex, and Annelie Rieger. 1985. "Medicine as Social Science: Rudolf Virchow on the Typhus Epidemic in Upper Silesia." *International Journal of Health Services* 15, no. 4: 547–559.

Thayer, Zaneta M., and Amy L. Non. 2016. "Anthropology Meets Epigenetics: Current and Future Directions." *American Anthropologist* 117, no. 4: 722–735.

The Packer. 2015. "Strawberry Growers Rev Up Bug Vacs." *The Packer*, April 2, 2015. https://www.thepacker.com/article/strawberry-growers-rev-bug-vacs.

Thomas, Madeleine. 2017. "Unknown, Unregulated, Undrinkable." *Pacific Standard*, June 14, 2017. https://psmag.com/news/unknown-unregulated-undrinkable#.bos7hubrv.

Tracking California. n.d. "Pesticides." Home page, accessed September 7, 2020. https://trackingcalifornia.org/pesticides/pesticides-landing.

TriCal, Inc. n.d. "TriCal, Inc. Company History." Web page, accessed March 12, 2020. https://www.trical.com/aboutus/companyhistory.html.

Tuhiwai Smith, Linda. 1999. *Decolonizing Methodologies. Research and Indigenous Peoples*. New York: Zed Books.

Tyner, James. 2019. *Dead Labor: Toward a Political Economy of Premature Death*. Minneapolis: University of Minnesota Press.

UC IPM (University of California Integrated Pest Management). 2018. "Non-Fumigant Alternatives for Soil Disinfestation." Web page, last updated July 2018. https://www2.ipm.ucanr.edu/agriculture/strawberry/Non-Fumigant-Alternatives-for-Soil-Disinfestation/.

Udall, Tom. 2018. "28 Senators to EPA: Don't Weaken Rules Protecting Workers & Children from Toxic Pesticides." Tom Udall for Senate, March 13, 2018. http://www.tomudall.senate.gov/news/press-releases/28-senators-to-epa-dont-weaken-rules-protecting-workers-and-children-from-toxic-pesticides.

Uesugi, Tak. 2016. "Toxic Epidemics: Agent Orange Sickness in Vietnam and the United States." *Medical Anthropology* 35, no. 6: 464–476.

Ulloa, Jazmine. 2017. "How California's Trust Act Shaped the Debate on the New 'Sanctuary State' Proposal." *Los Angeles Times*, September 10, 2017. https://www.latimes.com/politics/la-pol-ca-trust-act-sanctuary-state-immigration-20170910-htmlstory.html.

United Farm Workers. 2010. "UFW's National 'Take Our Jobs' Campaign Invites U.S. Citizens to Replace Immigrant Farm Workers." United Farm Workers, June 24, 2010. https://

ufw.org/UFW-s-national-Take-Our-Jobs-campaign-invites-U-S-citizens-to-replace
-immigrant-farm-workers/.

Unterberger, Alayne. 2018. "'No One Cares If You Can't Work!': Injured and Disabled Mexican-Origin Workers in Transnational Life Course Perspective." *Anthropology of Work Review* 39, no. 2: 105–115.

———. 2009. "The Blur: Balancing Applied Anthropology, Activism and Self vis-a-vis Immigrant Communities." *NAPA Bulletin* 31: 1–12.

U.S. GAO (U.S. Government Accountability Office). 2014. *Food Safety: FDA and USDA Should Strengthen Pesticide Residue Monitoring Programs and Further Disclose Monitoring Limitations.* GAO-15-38. Washington, DC: U.S. GAO. https://www.gao.gov/assets/670 /666408.pdf.

USDA ERS (United States Department of Agriculture, Economic Research Service). 2020. "Farm Labor." Web page, last updated April 22, 2020. https://www.ers.usda.gov/topics /farm-economy/farm-labor/#demographic.

Van den Bosch, Robert. 1989. *The Pesticide Conspiracy.* Berkeley: University of California Press.

Vandenberg, Laura N., Theo Colborn, Tyrone. B. Hayes, Jerold J. Heindel, David R. Jacobs, Duk-Hee Lee, Toshi Shioda, Ana M. Soto, Frederick S. vom Saal, Wade V. Welshons, R. Thomas Zoeller, and John Peterson Myers. 2012. "Hormones and Endocrine-Disrupting Chemicals: Low Dose Effects and Nonmonotonic Dose Responses." *Endocrine Reviews* 33, no. 3: 378–455.

Vanderhorst, Daniel. 2017. "Cal Giant Hosts Berry Planting Field Trip." *The Packer,* November 21, 2017. https://www.thepacker.com/article/cal-giant-hosts-berry-planting-field-trip.

Velmurugan, Ganesan, Tharmarajan Ramprasath, Mithieux Gilles, Krishnan Swaminathan, and Subbiah Ramasamy. 2017. "Gut Microbiota, Endocrine-Disrupting Chemicals and the Diabetes Epidemic." *Trends in Endocrinology and Metabolism* 28, no. 8: 612–625.

Veteto, James R., and Josh Lockyer. 2015. "Applying Anthropology to What? Tactical/Ethical Decisions in an Age of Global Neoliberal Imperialism." *Journal of Political Ecology* 22, no. 1: 357–367.

Vidrio, Edgar, Pamela Wofford, Randy Segawa, and Jay Schreider. 2014. *Air Monitoring Network Results for 2013.* Sacramento: California Environmental Protection Agency, Department of Pesticide Regulation, 2014.

Villarejo, Don, Stephen A. McCurdy, Bonnie Bade, Steve Samuels, David Lighthall, and Steve Williams III. 2010. "The Health of California's Immigrant Hired Farmworkers." *American Journal of Industrial Medicine* 54: 387–397.

Wark, Julie, and Daniel Raventós. 2018. *Against Charity.* Pretoria, South Africa: Counterpunch.

Warnert, Jeannette E. 2010a. "Soil Fumigant History." Farm Progress, May 28, 2010. https:// www.farmprogress.com/management/soil-fumigant-history.

———. 2010b. "UC Riverside Scientist Supports Methyl Iodide Registration." *ANR News Blog,* July 6, 2010. https://ucanr.edu/blogs/blogcore/postdetail.cfm?postnum=3024.

Warren, Kay B. 2006. "Perils and Promises of Engaged Anthropology: Historical Transitions and Ethnographic Dilemmas." In *Engaged Observer: Anthropology, Advocacy, and Activism,* edited by Victoria Sanford and Asale Angel-Ajani, 213–227. New Brunswick, NJ: Rutgers University Press.

Weeks, Samuel. 2015. "Longing for 'Normal' Post-Fordism: Cape Verdean Labor-Power on a Lisbon Periphery in Crisis." *Anthropology of Work Review* 36, no. 1: 13–25.

Weir, David, and Mark Schapiro. 1987. *Circle of Poison: Pesticides and People in a Hungry World.* San Francisco: Institute for Food and Development Policy.

Wells, Miriam J. 1996. *Strawberry Fields: Politics, Class, and Work in California Agriculture.* Ithaca, NY: Cornell University Press.

White, Timothy P. 2017. "Memorandum: Statement regarding Deferred Action on Childhood Arrivals." California State University, Office of the Chancellor, September 5, 2017. http://fresnostate.edu/advancement/ucomm/documents/messaging/Chancellor-White-Memo-9-5-17-DACA.PDF.

Whorton, James C. 1974. *Before Silent Spring: Pesticides & Public Health in Pre-DDT America*. Princeton, NJ: Princeton University Press.

Wilde, Mary H. 2003. "Embodied Knowledge in Chronic Illness and Injury." *Nursing Inquiry* 10, no. 3: 170–176.

Wiley, Andrea S. 2013. *Cultures of Milk: The Biology and Meaning of Dairy Products in the United States and India*. Cambridge, MA: Harvard University Press.

Wilhelm, Stephen, and James E. Sagen. 1974. *A History of the Strawberry: From Ancient Gardens to Modern Markets*. Berkeley: University of California, Division of Agricultural Sciences.

Williams, Brett. 2015. "Fragments of a Limited Mutuality." In *Mutuality: Anthropology's Changing Terms of Engagement*, edited by Roger Sanjek, 238–248. Philadelphia: University of Pennsylvania Press

———. 2001. "A River Runs through Us." *American Anthropologist* 103, no. 2: 409–431.

———. 1984. "Why Migrant Women Feed Their Husbands Tamales: Foodways as a Basis for a Revisionist View of Tejano Family Life." In *Ethnic and Regional Foodways in the United States: The Performance of Group Identity*, edited by Linda Keller Brown and Kay Mussell, 113–126. Knoxville: University of Tennessee Press.

———. 1975. "The Trip Takes Us: Chicano Migrants on the Prairie." PhD diss., University of Illinois at Urbana-Champaign.

Wong, Anselm, Rais Vohra, Anne-Michelle Ruha, Zeff Koutsogiannis, Kimberlie Graeme, Paul I. Dargan, David M. Wood, and Shaun L. Greene. 2015. "The Global Educational Toxicology Uniting Project (GETUP): An Analysis of the First Year of a Novel Toxicology Education Project." *Journal of Medical Toxicology* 11, no. 3: 295–300.

World Health Organization. 1990. *Public Health Impact of Pesticides Used in Agriculture*. Geneva: World Health Organization, 1990. https://apps.who.int/iris/handle/10665/39772.

Wright, Angus. 2005. *The Death of Ramón González: The Modern Agricultural Dilemma*. Austin: University of Texas Press.

Wylie, Sara Ann. 2018. *Fractivism: Corporate Bodies and Chemical Bonds*. Durham, NC: Duke University Press.

Yeung, Bernice, and Grace Rubenstein. 2013. "Rape in the Fields: Female Workers Face Rape, Harassment in U.S. Agriculture Industry." *PBS Frontline*, June 25, 2013. https://www.pbs.org/wgbh/frontline/article/female-workers-face-rape-harassment-in-u-s-agriculture-industry/.

Zavella, Patricia. 2016. "Mexicans' Quotidian Struggles with Migration and Poverty." In *The New Latino Studies Reader: A Twenty-First Century Perspective*, edited by Ramón A. Gutiérrez and Tomás Almaguer, 235–265. Berkeley: University of California Press.

———. 2011. *I'm Neither Here nor There: Mexicans' Quotidian Struggles with Migration and Poverty*. Durham, NC: Duke University Press.

———. 2001. "The Tables Are Turned: Immigration, Poverty, and Social Conflict in California Communities." In *New Poverty Studies: The Ethnography of Power, Politics, and Impoverished People in the United States*, edited by Judith Goode and Jeff Maskovsky, 103–133. New York: New York University Press.

Zham, Shelia Hoar, and Mary H. Ward. 1998. "Pesticides and Childhood Cancer." *Environmental Health Perspectives* 106, suppl. 3: 893–908.

Zlolniski, Christian. 2019. *Made in Baja: The Lives of Farmworkers and Growers behind Mexico's Transnational Agricultural Boom*. Berkeley: University of California Press.

INDEX

Note: Page numbers in italics denote an illustration; an "n" after a page number denotes a note

ABOUT THE AUTHOR

DVERA I. SAXTON is an associate professor of anthropology at California State University–Fresno, where she teaches introductory courses and does creative project-based learning with medical and environmental anthropology undergraduate students. She completed her PhD in anthropology, with an emphasis in race, gender, and social justice, at the American University in Washington, DC. For 2013–2014 she was a postdoctoral fellow with the Social Science Environmental Health Research Institute at Northeastern University in Boston. In California's Central Valley, Saxton remains active in and outside the classroom on issues related to community health, agriculture, and food and environmental justice. In 2020, she became a Senior Atlantic Fellow for Equity in Brain Health with the Global Brain Health Institute.